Isaac R. Butts

The Tinman's Manual

And builder's and mechanic's handbook, designed for tinmen, japanners,

coppersmiths, engineers. Third Edition

Isaac R. Butts

The Tinman's Manual
And builder's and mechanic's handbook, designed for tinmen, japanners, coppersmiths, engineers. Third Edition

ISBN/EAN: 9783337184261

Printed in Europe, USA, Canada, Australia, Japan

Cover: Foto ©berggeist007 / pixelio.de

More available books at **www.hansebooks.com**

THE

TINMAN'S MANUAL

AND

BUILDER'S

AND

MECHANIC'S HANDBOOK,

DESIGNED FOR

Tinmen, Japanners, Coppersmiths, Engineers, Mechanics, Builders, Millwrights, Smiths, Masons, Carpenters, Joiners, Slaters, Plasterers, Painters, Glaziers, Pavers, Plumbers, Surveyors, Gaugers, &c., &c.; with Compositions and Receipts for other useful and important purposes in the Practical Arts.

By I. R. BUTTS,

Author of the "United States Business Man's Law Cabinet," "Business Man's Law Library;" "Merchant's and Shipmaster's Manual and Shipbuilder's and Sailmaker's Assistant," &c., &c.

THIRD EDITION.

BOSTON:

PUBLISHED BY I. R. BUTTS & CO.

CORNER OF SCHOOL AND WASHINGTON STREET,

Over Ticknor & Fields' Bookstore.

1863.

PREFACE.

THE present work is offered to Tinmen, Builders, Mechanics, and Engineers, as a useful manual of reference, and information.

The first part of the work containing RULES, DIAGRAMS and TABLES, will be found very useful to Tinmen.

Mr. Truesdell who has, for many years, used the Diagrams prepared by him for this work, now offers them to the public with every confidence.

The Receipts for Japans, Varnishes, Cements, &c., were taken from "Ure's Dictionary," "Cooley's Cyclopedia," "Muspratt's Chemistry," and other valuable publications.

The sources from which most of the materials relating to *Building*, *Mechanics*, and *Engineering* have been derived, are "Grier's Mechanic's Calculator," "Templeton's Workshop Companion," "The Engineer's and Contractor's Pocket-book," "Adcock's Engineer," "Smeaton's Builder's Companion," and "Lowndes's Engineer's Handbook," *which renders this portion of the work deserving of the utmost confidence.*

LETTER FROM L. W. TRUESDELL.

Mr. Butts,—

DEAR SIR,—If I may be permitted to comment upon the first part of your book, I would like to point out to Tinmen the value of the Diagrams which, a few years ago, could not have been purchased at any price ; but as they are now to be published, and sold at a low price, I am confident they will be bought by every Tinman, for I know, by experience, the perplexities to which they are often subjected from the want of them.

With these Directions and Diagrams, the Tinman will be enabled to cut a Right-Angled or Circular Elbow of any size, in a few minutes, and produce as perfect a mitre joint as can be made ; also, patterns for Flaring vessels, of any size or flare, Envelopes for Cones, Pyramid Cakes, Covers for Oval Dishes and Boilers, Funnel-shaped Covers for Pails, Breasts for Cans, Lips for Measures of any size, &c.*

When about to make a copy from these diagrams the person should provide himself with a sheet of paper or tin-plate, and strictly follow the directions given.

Suppose, for example, that he is about to copy Fig. 1, the directions are, *first,* from the centre C describe a circle AB. Having described the circle AB, next, place the corner of the square on the centre C, and draw the lines CD and CE ; then draw the chord DE.

When the Tinman has become familiar with the diagrams, he will find them simple and convenient, and be better qualified to undertake work of a difficult character. If an Elbow at right-angles, of ten or fifteen inches diameter, should be required, with the directions and diagrams before him, he could cut it out in a few minutes ; and so with a curved elbow of any diameter, a semicircle, or an ellipses-shaped dish of any size. But without a rule or pattern it would be a difficult and troublesome undertaking.

Having by experience proved the correctness and usefulness of these Diagrams, I can confidently recommend them to all persons engaged in the manufacture of Tin Ware.

L. W. TRUESDELL.

Owego, N. Y. Sept. 23, 1860.

EXTRACT OF A LETTER FROM A TINMAN.

Mr. Butts,—

Dear Sir, — " Your ' *Tinman's Manual* ' strikes me as being nearer what we want in our business, than anything I have ever seen,—and I have examined every thing of the kind I have been able to find. The best we have been able to do has been to pick up what ideas we could from works on Geometry and Building, and work out what rules we could from them. I have often wondered why some person did not undertake just what you have done. This work of yours supplies just the want that every thinking man who works at the business has felt, even from his first start ; and the want is still more sensibly felt as he grows older, and finds how much there is to learn.''

---

* In Tinman's Diagrams the allowance for locks is always omitted.

CONTENTS.

RULES AND DIAGRAMS FOR WORKERS IN TIN, SHEET IRON AND COPPER.

Page.

Manufacture of Tin Plate........ 12
Quality of Tin Plate............ 14

CIRCLES.

To find the Circumference of any
    Diameter 15
To find the Area of a Sector of a
    Circle....................... 15
Proportion of Circles to enable ma-
    chinists to enlarge or reduce
    wheels without changing their
    motion 16
The Circle and its Sections....... 27
To find the centre of a Circle from
    a part of the Circumference..... 33
Diameters, Circumferences, and
    Areas of Circles............... 41

CYLINDERS.

To find the Contents in Gallons of
    any Cylindrical Vessel........ 38
Tables giving the Content in Gal-
    lons of Cylinders from 1 inch to
    30 feet Diameter............... 42
Table giving the Content in Gal-
    lons of Cans from 3 inches to 40
    inches Diameter............... 45

BEVEL COVERS.

To describe Bevel Covers for Ves-
    sels, or Breasts for Cans 25
To describe Bevel Covers for Ves-
    sels, or Breasts for Cans, (another
    mode)........ 32
To describe Covers for Pails...... 25

ELLIPSES OR OVALS.

To describe an Ellipse........... 17
Definition of an Oval,—note....... 17
To describe an Ellipse (another
    mode) 18
To find the Circumference of an
    Ellipse 19
To find the Area of an Ellipse..... 19
To describe an Oval Boiler Cover 26
To draw an Ellipse, the transverse
    and conjugate Diameters being
    given, i. e. the length and width 116
To draw an Ellipse by means of
    two concentric circles......... 117

Page.

ELBOWS.

To describe a Right Angled Elbow 20
To describe a Straight Elbow (old
    method)...................... 21
To describe a Curved Elbow..... 22
To describe a Straight Elbow
    (another mode) 24

FLARING VESSELS.

To describe a Flaring Vessel Pat-
    tern, a Set of Patterns for a Py-
    ramid Cake, or an Envelope for
    a Cone...................... 28
To describe a Cone or Frustum... 29
To strike the Side of a Flaring
    Vessel...................... 31
To construct the Frustum of a Cone 34
To strike out a Cone or Frustum.. 35
To find the content of a Cone ... 35
To find the Angles of a Frustum of
    an inverted Pyramid, such as a
    Mill Hopper, &c............... 36
To find the content of the Frustum
    of a Cone, such as a Coffee-pot,
    Bowl, &c.................... 36

MISCELLANEOUS.

To joint Lead Plates............ 23
Soldering for Lead, Zinc, Tin, and
    Pewter..................... 23
To joint Lead Pipes,............ 24
Soldering for Copper............ 160
To describe a Lip to a Measure.. 27
To describe a Cycloid, or Curve.. 30
To describe a Heart............. 30
Tinning Iron................... 31
A good Solder.................. 32
Sector, for obtaining Angles...... 34
Sector, definition of............. 34
Rule to find the Content in Gallons
    of Frustums of Cones......... 37
Rule to find the Content in Gallons
    of any Cylindrical Vessel....... 38
Table to ascertain the weight of
    Pipes of various metals, and any
    Diameter required............. 38
Table of Tin Plates, size and
    weight per box 39
Table of Cans, quantity and qual-
    ity of Tin required for 2½ to 125
    gallons.................. ... 39

6   CONTENTS.

	Page.		Page.
Weight of a cylindrical and cubic inch, cubic foot and gallon of Water...	40	Crystallizing Tin Plate, how performed...	46,
Decimal Equivalents to the fractional parts of a Gallon or an Inch	40	Tinning Vessels of Brass or Copper	46
Tables containing the Diameters, Circumferences and Areas of Circles...	42	Kustitien's Metal for Tinning...	46
		Instruments used in Drawing...	101
Tables giving the Diameters and Circumferences of Circles...	171	Composition of Britannia Metal for Spouts, Registers, Spoons, &c..	91
Tables to ascertain the weight of Lead Pipes...	139	Composition of Britannia Metal for Lamps, Pillars, Handles, and Castings...	92
Capacity of Cans in Gallons from 3 inches to 40 inches in Diameter	45	Solder for Britannia Ware...	91
New Tinning Process...	46	Lacker for Tin Plate...	73 & 94
		Solder, Tinman's...	96
		Definitions of Arithmetical Signs used in this work...	110

RECEIPTS FOR THE USE OF JAPANNERS, VARNISHERS, BUILDERS, MECHANICS, &c.

JAPANNING AND VARNISHING.

Directions for Japanning...	49	Soft Brilliant Varnish...	62
White Japan Grounds—Gum Copal	50	Brown Hard Spirit Varnishes—To prepare a Varnish for Coating Metals — Varnish for Iron and Steel, for Iron Work, Black for Iron Work, Bronze for Statuary	63
Black Grounds—Black Japan....	51	
Brunswick Black — Blue Japan Grounds—Scarlet Japan—Yellow Grounds — Green Japan Grounds...	52	
Orange Colored Grounds—Purple Japan Grounds — Black Japan—Japan Black for Leather—Transparent Japan—Japanners' Copal Varnish...	53	Amber Varnishes, Black, Pale, Hard—Black Varnish...	64
		Varnish for certain parts of Carriages, Coaches, Mahogany, for Cabinet Makers—Cement Varnish for water-tight Luting—The Varnish of Watin for Gilded Articles — Oak Varnish —Varnish for Wood-work—Dark Varnish for light Wood-work...	65
Tortoise Shell Japan — Painting Japan Work — Japanning Old Tea-trays—Japan Finishing....	54	

VARNISHES—MISCELLANEOUS.

Substances employed for making Varnishes...	55	Varnish for Instruments, for Wood Toys of Spa, for Furniture—To French Polish...	66
Choice of Linseed Oil. ...	56	Furniture Polishes, Gloss, Cream, Oils, Pastes—Etching Varnishes	67

CHIEF RESINS EMPLOYED IN MAKING VARNISH.

Amber—Anime—Benzoin — Colophony—Copal...	56	Varnish for Engravings, Maps, to fix Engravings or Lithographs on Wood, for Oil Paintings and Lithographs, for Paintings and Pictures—Milk of Wax...	68
Dammara—Elimi—Lac—Mastic—Sandarach...	57	
Turpentine — Alcohol — Naphtha and Methylated Spirit of Wine—Spirit Varnishes...	58	Crystal Varnishes, Italian—Water Varnish for Oil Paintings—Varnish for Paper-hangings, Bookbinders, Cardwork...	69
Essence Varnishes—Oil Varnishes —Lacker...	59	Varnish for Printers — for Brick walls—Mastic Varnishes—India Rubber Varnishes...	70

VARNISHES.

Copal Varnishes (six kinds)...	60	Black Varnish for Harness—Boiled Oil or Linseed Oil Varnish—Dammar Varnish...	71
Copal Varnishes (three kinds) Cabinet Varnish—Table Varnish—Common Table Varnish—Copal Varnish for Inside Work...	61	Common Varnish — Waterproof Varnishes — Varnishes for Balloons, Gas Bags, &c.—Gold Varnish — Wainscot Varnish for House Painting and Japanning	72
Copal Polish—White Spirit Varnish—White Hard Spirit Varnishes—White Varnish...	62	

LACKERS.

Gold Lacker—Red Spirit Lacker—Pale Brass Lacker—Lacker for

Page.

Tin — Lacker Varnish — Deep Gold Colored Lacker—Lackers for Pictures, Metal, Wood, or Leather...................... 73

CEMENTS.

Armenian, or Diamond Cement.. 74
Cements for mending Glass Ware 74
Cement for Stone-ware--Iron-Rust Cement—for making Architectural Ornaments—Varley's Mastic —Electrical and Chemical Apparatus Cement.................... 75
Cements for Iron Tubes, Boilers, Ivory, Mother of Pearl, Holes in Castings, Coppersmiths and Engineers, Plumbers, Bottle corks, China and Leather............... 76
Cements for Marble, Marble-workers, Coppersmiths, Glass, mending Iron Pots and Pans, Cisterns and Casks................... 77
Cements for mending Fractured Bodies of all kinds, for Cracks in Wood, joining Metals and Wood, for fastening Brass to Glass Vessels, Blades, and Files--Gas-Fitter's Cement—Cement Paint.... 78

BUILDERS' CEMENTS.

Cements for Terraces, Roofs, Reservoirs, Fronts of Houses, &c.. 79
Cements for Brick Walls, Seams, and Tile roofs................. 80
Coarse Stuff...................... 80
Parker's Cement—Hamelein's Cement—Plaster in imitation of Marble—Scagliola............. 81
Maltha, or Greek Mastic — Fine Stuff—Stucco for Inside Walls 82
Higgins's Stucco — Gauge Stuff—

Page.

Composition — Foundations of Buildings.................... 83
Concrete Floors—Fire-proof Composition...................... 84

RECEIPTS.

To Polish Wainscot and Mahogany—Imitation of Mahogany—Furniture Varnish — To make Glass and Stone Paper........ 85
Whitewash — Paint for Coating Wire Work—To Bleach Sponge —Lac Varnish for Vines— Razor Paste — Leather Varnish — To keep Tires Tight on Wheels.... 86
To Cut Glass — Prepared Liquid Glue—Marine Glue— Paste for Envelopes—Dextrine, or British Gum—Gum Mucilage......... 87
Flour Paste — Sealing Wax for Fruit Cans—Fusible Metal--Metallic Cement.................. 88
Artificial Gold—Or-mulo—Blanched Copper—Browning Gun Barrels—Silvering Powder for Coating Copper.................... 89
Alloys for Journal Boxes—Bells of Clocks—Tools—Cymbals and Gongs—Solder for Steel Joints—Files—To prevent Tools from Rusting—Axle-Grease—to Galvanize—Soft Gold Solder....... 90

RECEIPTS AND COMPOSITIONS.

Nearly 200 Compositions for Mechanists, Iron and Brass Founders, Turners, Tinmen, Coppersmiths, Dentists, Finishers of Brass, German Silver, Britannia, and other useful purposes in the Practical Arts............. 91

MECHANICAL DRAWING.

Instruments used in Drawing.... 101
The Sector 103
Mechanical Drawing and Perspective............................. 105

PRACTICAL GEOMETRY.

Definition of Arithmetical Signs.. 110

PROBLEMS.

To find the Circumference of a Diameter........................ 15
To find the area of a Sector...... 15
To find the Proportion of Circles by which to enlarge or reduce Wheels without changing their motion..................... 16
To find the various and proper Dimensions of Materials whereby to construct Hipped Roofs,&c... 36

To find the Centre of a Circle from a part of the Circumference...... 33
The Circle and its Sections....... 27
Sector, for obtaining Angles.....: 34
To inscribe an Equilateral Triangle within a given Circle....... 111
Within a given Circle to inscribe a Square 112
Within a given Circle to inscribe a regular Pentagon............. 112
Within a given Circle to describe a regular Hexagon............. 113
To cut off the Corners of a given

Page.

Square, so as to form a regular Octagon.................... 113
To divide a given Line into any Number of Parts, which Parts shall be in the same Proportion to each other as the Parts of some other given line, whether those parts are equal or unequal 114
On a given Line to draw a Polygon of any Number of Sides, so that that Line shall be one side of a Polygon.................... 114

OF DRAWING CURVED LINES.

To draw an Ellipse with the Rule and Compasses, the transverse and conjugate Diameters being given ; i. e. the length and width 116
To draw an Ellipse by means of two Concentric Circles....-.... 116
To draw an Ellipse of any length and width...................... 18
To find the Circumference & Area of an Ellipse.................... 19
Other methods for describing an Ellipse...................... 117
To find the Centre and the two Axes of an Ellipse............. 118
To draw a flat Arch by the intersection of Lines, having the Opening and Spring or Rise given..................... 119
To find the Form or Curvature of a raking Moulding that shall unite correctly with a level one 119
To find the Form or Curvature of the Return in an open or broken Pediment..................... 120

EPITOME OF MENSURATION.

Of the Circle, Cylinder, Sphere, Zone, &c.................... 122
Of the Square, Rectangle, Cube 123
Surfaces and solidities of Bodies 124
Of Triangles, Polygons, &c........124
Of Ellipses, Cones, Frustums, &c. 125

INSTRUMENTAL ARITHMETIC.

Utility of the Slide Rule........... 125
Numeration..................... 126

To Multiply Numbers by the Rule 126
To divide Numbers upon the Rule 126
Proportion or Rule of Three Direct 127
Square & Cube Roots of Numbers 127
Rule of Three Inverse........... 127
Mensuration of Surface........... 128
Mensuration of Solidity and Capacity 129
Power of Steam Engines........ 130
Of Engine Boilers.............. 130

RULES AND TABLES FOR ARTIFICERS AND ENGINEERS.

Measurement of Bricklayer's work 132
Table to find the number of Bricks in any given Wall.............. 133
Measurement of Wells & Cisterns 133
Measurement of Mason's Work.. 133
Measurement of Carpenter's and Joiner's Work................ 134
Table of different sized Nails to a lb 135
Table of different sized Sashes, &c 136
Measurement of Slater's Work... 136
Table of American Slates......... 136
Table of Imported Slates.......... 137
Measurement of Plasterer's Work 137
Measurement of Paver's Work.... 137
Measurement of Painter's Work... 137
Measurement of Glazier's Work.. 138
Table of Size and Number of Lights to the 100 Square Feet... 138
Measurement of Plumber's Work 138
Table of Sizes and Weight of Patent Lead Pipe................. 139
Table of Boston Lead Pipe....... 139
Table of Comparative Strength and Weight of Ropes and Chains ... 139

STRENGTH OF MATERIALS.

Definitions..................... 140
Table of Tenacities, Resistance to

Compression, &c., of various Bodies .....?............... 140
Resistance to Lateral Pressure... 140
Table of Practical Data.......... 141
To find the dimensions of a beam of Timber to sustain a given Weight..................... 141
To determine the absolute strength of a Rectangular Beam of Timber 141
To determine the dimensions of a Beam with a given degree of deflection..................... 142
Cast-iron Beams of strongest section.................... 142
Of Wooden Beams, Trussed....... 142
Absolute Strength of Cast-iron Beams...................... 142
Dimensions for Cast-iron Beams.. 143
To find the Weight of a Cast-iron Beam·..................... 143
Resistance to flexure by vertical pressure.................... 143
To determine the dimensions for a Column of Timber.............. 144
Resistance of Bodies to Twisting 144
Relative strength of Metals to resist Torsion.................. 144

Page.

Breaking strength of a Bar of Wrought Iron.................... 145
Lateral strength of Wrought Iron as compared with Cast-iron..... 145
Load on Bridges, Floors, Roofs, and Beams 145
Strength of Beams, Bar of Wood, Stone, Metal, Ropes, Tubes, or Hollow Cylinders.............. 146
Models proportioned to Machines 146
Metals arranged according to their Strength....................... 147
Woods arranged according to do. 147
Strength of Cords, &c............ 147
Strength of Rectangular and Round Timber........................ 148
Table of the Cohesive Power of Bars of Metal.................. 148
Relative Strength of Cast and Malleable Iron.................... 148

STRENGTH OF BEAMS.

Solid, Rectangular, Round, Hollow 149
To find the breaking Weight in lbs. 149
To find the proper Size for any given purpose.................... 150
Strength of Cast-iron with Feathers or Flanges................. 150
Wrought Iron Beams and Girders 151
Hollow Girders.................. 152
To find the Strength of a Round Girder....................... 152
To find the Strength of any Beam 152

SOLID COLUMNS.

To find the Strength of any Wro't Iron Column with Square ends 153
To find the Strength of Round Columns exceeding 25 diameters in Length....................... 154
Tables of Powers for the Diameters and Lengths of Columns... 154

HOLLOW COLUMNS.

Square Columns of Plate Iron rivetted.......................... 155
To find the Strength of any Hollow Wrought Iron Column 155
Round Columns of Plate Iron 156

CRANE.

To find the Strain on the Post.... 156

COLD WATER PUMP.

To find the proper Size, under any circumstances, capable of supplying twice the quantity ordinarily used in injection.......... 156

FANS.

Velocity of Fans.................. 157
The best Velocity of Circumference for different Densities.... 157

Page.

To find the Horse Power required for any Fan.'.................. 157
To find the Density to be attained with any given Fan............ 157
To find the Quantity of Air that will be delivered by any Fan, the Density being known....... 158

FRICTION.

From Mr. Rennie's Experiments.. 158

CENTRIFUGAL FORCE

In terms of Weight.............'.... 158

PEDESTAL AND BRACKET.

Thickness of cover, diameter, distance, solid metal, &c.......... 159

TEMPERING.

For Lancets, Razors, Penknives, Scissors, Hatchets, Saws, Chisels, Springs, &c.............. 159

CASE HARDENING

Articles, how Case Hardened.... 159
To Case Harden Cast Iron....... 160

HEAT.

Effects of Heat on Metals, &c., at certain Temperatures.......... 160

SOLDERING.

For Joints, Copper, Iron and Brass 160

BORING.

The best speed for boring Iron, drilling, and turning............ 161

BRASS.

Compositions of Brass.......... 161
Brass Castings, mode of Casting.. 161

ROPE.

To find the Breaking Weight of Tarred Hemp Rope............ 162
To find the Weight per Fathom of Rope or Tarred Cordage....... 163
To find the Weight per Fathom of Tarred Hawser or Manilla Rope 163
To find the Weight per Fathom of Hawser laid Manilla.......... 163

WEIGHT OF CASTINGS.

To find the Weight of any Casting 163
To find the Weight from the Areas 163
To find the Weight in cwts....... 163
Weight of Boiler Plates.......... 163
To find the Weight of Boiler Plates 164

CONTINUOUS CIRCULAR MOTION.

When Time is not taken into Account....................... 16

Page.

To find the number of Revolutions of the last to one of the first, in a train of Wheels and Pinions.... 164

When Time must be regarded.... 165

The distance between the Centres and Velocities of two Wheels being given, to find their Diameters 165

To determine the Proportion of Wheels for Screw-cutting by a Lathe........................ 166

Table of Change Wheels for Screw-cutting; the leading Screw being half inch pitch, or containing 2 threads in an inch........ 167

Table by which to determine the Number of Teeth, or Pitch of Small Wheels, or what is called the Manchester Principle....... 167

Strength of the Teeth of Cast Iron Wheels at a given Velocity..... 168

WHEELS AND GUDGEONS.

To find size of Teeth necessary to transmit a given Horse Power... 168

To find the Horse Power that any Wheel will transmit.......... 169

Page.

To find the multiplying Number for any Wheel.................... 169

To find the Size of Teeth to carry a given Load in lbs............ 169

WATER.

To find the Quantity of Water that will be discharged through an Orifice, or Pipe, in the side or bottom of a Vessel............ 169

To find the size of Hole necessary to discharge a given Quantity of Water under a given Head..... 170

To find the Height necessary to discharge a given Quantity thro' a given Orifice................ 170

The Velocity of Water issuing from an Orifice in the side or bottom of a Vessel ascertained.... 170

To find the Quantity of Water that will run through any Orifice, the top of which is level with the Surface of Water, as over a Sluice or Dam.................. 170

To find the Time in which a Vessel will empty itself through a given Orifice...................... 170

MECHANICAL TABLES FOR THE USE OF OPERATIVE SMITHS, MILLWRIGHTS, AND ENGINEERS.

Tables of the Diameters and Circumferences of Circles.......... 171

Observations on do.............. 177

Circumferences of Angled Iron Hoops—outside................ 179

Circumferences of Angled Iron Hoops—inside.................. 180

Observations on the above Tables 181

Tables of the Weight of 100 lbs. of Ship Spikes, Hatch Nails, Hook Heads, Deck Nails, Boat Spikes, Railroad Spikes & Horse Shoes 182

Coppers, dimensions and weight of 183

Copper Tubing, weight of........ 183

Brass, Copper, Steel and Lead, weight of a Foot from ¼ to 3 inches Round or Square............ 183

Flat Cast Iron, weight of a Foot... 184

Cast Iron, Weight of a Superficial Foot, from ¼ to 2 inches thick.. 184

Table giving the Weight of Cast Iron, Copper, Brass, and Lead Balls, from 1 to 12 inch diameter 184

Cast Iron, weight of a Foot in length of Square and Round.... 185

Steel, weight of a Foot of Flat..... 185

Parallel Angle Iron, of equal sides 186

Parallel Angle Iron, unequal sides 186

Taper Angle Iron, of equal sides.. 186

Parallel T Iron, unequal width and depth........................ 187

Parallel T Iron, of equal depth and width........................ 187

Taper T Iron..................... 187

Table of Weight of Sash Iron.... 188

Table of Weight of Rails, top and bottom Tables.................. 188

Table of Weight of Temporary do. 188

Tables showing the Weight of a lineal Foot of Malleable Rectangular, or Flat Iron, from ⅛ to 3 inches in thickness............ 190

ELASTIC FORCE OF STEAM.

Table of the Elastic Properties of Steam and corresponding temperature of Water................ 194

Production & Properties of Steam 195

Table of the Elastic Force of Steam the Pressure of the Atmosphere not being included.............. 195

Table of the Consumption of Coal per hour in Steamers.......... 196

Evaporative Power of Coal...... 196

GAUGER'S RULES AND TABLES.

To Gauge Casks, U. States Gallons 201

To Gauge Casks, Imperial Gallons 202

To Ullage, or find the contents of Casks partly filled.............. 203

Tables of the Comparative Value of Imperial and United States Measures..... 203

Miscellaneous Tables............. 204

RULES WITH DIAGRAMS

FOR WORKERS IN

TIN, SHEET IRON AND COPPER,

AND

TABLES GIVING THE DIAMETERS, CIRCUMFERENCES, AND AREAS OF CIRCLES,.

AND

THE CONTENTS OF EACH IN GALLONS.

MANUFACTURE OF TIN PLATE.

"The different processes of the manufacture of tin plate may be described most properly in seven distinct stages. The first begins with the bars of iron which form the plate ; the last terminates with an account of the process of tinning their surface. The description is somewhat technical ; but a glance at the following heads will enable the reader to comprehend the whole process :—

"1. *Rolling* is the first and most important point requisite to the production of the *latten*, or plates of iron, previous to the operation of tinning them. For this purpose the finest quality of charcoal iron is invariably employed, which, in its commercial state, generally consists of long flat bars. These are cut into small squares averaging one-half an inch in thickness, which are heated repeatedly in a furnace, and are repeatedly passing through iron rollers. A convenient degree of thinness having been obtained, the now extended plates are "doubled up," heated, rolled, opened-out, heated and rolled again, until, at length, the standard thickness of the plate has been reached.

"2. *Shearing.*—A pair of massive shears worked by machinery, is now applied to the rugged edges of this lamellar formation of iron-plate. It is cut into oblong squares, 14 inches by 10, and presents the appearance of a single plate of iron, beautifully smooth on its surface. A juvenile with a knife soon destroys the appearance, however, and eight plates are produced from the slightly coherent mass.

"3. *Scaling.*—This process consists in freeing the iron surface from its oxyd and scoriæ. After an application of sulphuric acid, a number of plates, to the extent, we shall say, of 600 or 800, are packed in a cast-iron box, which is exposed for some hours to the heat of a furnace. On being opened the plates are found to have acquired a bright blue steel tint, and to be free from surface impurities.

"4. *Cold Rolling.*—It is impossible that the plates could pass through the last fiery ordeal without becoming disfigured. The cold rolling process corrects this. Each plate is separately passed through a pair of hard polished rollers, screwed tightly together. Not only do the plates acquire from this operation a high degree of smoothness and regularity, but they likewise acquire the peculiar elasticity of hammered metal. One man will cold roll 225,000 plates in a week, and each of them is, on an average, three times passed through the rollers.

" 5. *Annealing.*—This process is also a modern improvement on the manufacture: 600 plates are again packed into cast iron boxes and exposed to the furnace. There is this difference in the present process from that of scaling—that the boxes must be preserved air-tight, otherwise the contained plates would inevitably weld together and produce a solid mass. The infinitessimal portion of confined air prevents this.

" 6. *Pickling.*—The plates are again consigned to a bath of diluted acid, till the surface becomes uniformly bright and clean. Some nice manipulation belongs to this process. Each plate is, on its removal from the acid, subjected to a rigid scrutiny by women, whose vocation it is to detect any remaining impurity, and scour it from the surface. These multifarious operations, it will be seen, are all preliminary to the last, and the most important of all—that of tinning. Theoretically simple, this process is practically difficult ; and to do it full justice would carry us beyond our limits. We shall however, mention the principal features.

" 7. *Tinning.*—A rectangular cast iron bath, heated from below, and calculated to contain 200 or 300 sheets, and about a tun of pure block tin, is now put in request. A stratum of pyreumatic fat floats upon its surface. Close to the side of this tin pot stands another receptacle, which is filled with melted grease, and contains the prepared plates. On the other side is an empty pot, with a grating ; and last of all there is yet another pot, containing a small stratum of melted tin. Let us follow the progress of a single plate. A functionary known as the " washerman," armed with tongs and a hempen brush, withdraws the plate from the bath of tin wherein it has been soaking ; and, with a degree of dexterity only to be acquired by long practice, sweeps one side of the plate clean, and then reversing it, repeats the operation. In an instant it is again submerged in the liquid tin, and is then as quickly transferred to the liquid grease. The peculiar use of the hot grease consists in the property it possesses of equalizing the distribution of the tin, of retaining the superfluous metal, and of spreading the remainder equally on the surface of the iron. Still there is left on the plate what we may term a salvage ; and this is

2

finally removed by means of the last tin pot, which just contains the
necessary quantity of fluid metal to melt it off—a smart blow being
given at the same moment to assist the disengagement. The "list-
mark," may be observed upon every tin plate without exception.
We may add here, that an expert washerman will finish 6000 metallic plates in twelve hours, notwithstanding that each plate is twice
washed on both sides, and twice dipped into the melted tin. After
some intermediate operations—for we need not continue the consec-
utive description—the plates are sent to the final operation of cleaning. For this purpose they are rubbed with bran, and dusted upon
tables ; after which they present the beautiful silvery appearance so
characteristic of the best English tin plate. Last of all they reach
an individual called the " sorter," who subjects every plate to a
strict examination, rejects those which are found to be defective, and
sends those which are approved to be packed, 300 at a time, in the
rough wooden boxes, with the cabalistic signs with which the most of
us have been familiar since the days of our adventures in the back-
shop of the tinsmith."—[*From the Builder.*]

QUALITY OF TIN PLATE.

The tests for tin plates are ductility, strength, and color ; and to
possess these, the iron used must be of the best quality, and all the
process be conducted with care and skill. The following conditions
are inserted in some specifications, and will serve to indicate the
strength and ductility of first-class tin plates : —

1st, They must bear cutting into strips of a width equal to ten
times the thickness of the plate, both with and across the fibre, without splitting ; the strips must bear, while hot, being bent upon a
mould, to a sweep equal to four times the width of the strip.

2nd, While cold, the plates must bear bending in a heading machine, in such a manner as to form a cylinder, the diameter of which
shall at most be equal to sixty times the thickness of the plate. In
these tests, the plate must show neither flaw nor crack of any kind

Explanation of Diagrams.

TO FIND THE CIRCUMFERENCE OF ANY DIAMETER.

[Drawn for this work by L. W. TRUESDELL, Tinman, Owego, N. Y.]

FIG. 1.

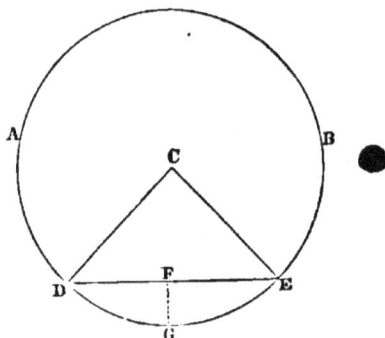

. From the centre C describe a circle AB, having the required diameter ; then place the corner of the square at the centre C, and draw the lines CD and CE ; then draw the chord DE : three times the diameter added to the distance from the middle of the chord DFE to the middle of the subtending arc DGE, will be the circumference sought.

TO FIND THE AREA OF THE SECTOR OF A CIRCLE.

RULE. Multiply the length of the arc DGE by its radius DC, and half the product is the area.

The length of the arc DGE equal 9½ feet, and the radii CD, CE, equal 7 feet required the area.

$$9 \cdot 5 \times 7 = 66 \cdot 5 \div 2 = 33 \cdot 25 \text{ the area.}$$

PROPORTION OF CIRCLES.

[Drawn for this work by L. W. TRUKSDELL, Tinman, Owego, N. Y.

Original.

FIG. 2.

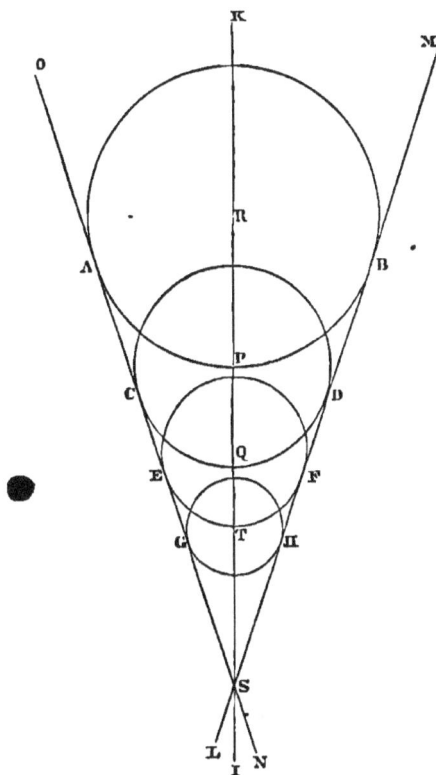

To enable machinists to enlarge or reduce machinery wheels without changing their respective motion.

First, describe two circles AB and CD the size of the largest wheels which you wish to change to a large or small machine, with the centre P of the smaller circle CD on the circumference of the large one AB ; then draw two lines LM and NO tangent to the circles AB and CD, and a line IK passing through their centres P and R ; then if you wish to reduce the machine, describe a circle the size you wish to reduce it to ; if one-half, for example, have the centre Q one-half the distance from R to S and describe the circle EF, and on its circumference T as a centre, describe a circle GH, allowing their circumferences to touch the tangent lines LM and NO, which will make the circle EF one-half the size of the circle AB, and GH one-half the size of CD ; therefore EF and GH are in the same proportion to each other as AB and CD.

If you wish to reduce one-third, have the centre Q one-third the distance from R to S ; if one-fourth have the centre Q one-fourth the distance from R to S, and so on. This calculation may be applied beyond the centre R for enlarging machine wheels, which will enable you to make the alteration without changing their respective motion.

TO DESCRIBE AN ELLIPSE, or OVAL.
[Simple Method.]
FIG. 3.

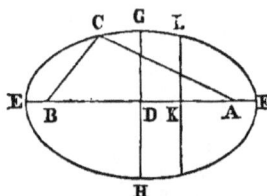

At a given distance, equal to the required eccentricity of the ellipse, place two pins, A and B, and pass a string, ACB, round them ; keep the string stretched by a pencil or tracer, C, and move the pencil along, keeping the string all the while equally tense, then will the ellipse CGLFH be described. A and B are the foci of the ellipse, D the centre, DA or DB the eccentricity, EF the principal axis or longer diameter, G H the shorter diameter, and if from any point L in the curve a line be drawn perpendicular to the axis, then will LK be an ordinate to the axis corresponding to the point L, and the parts of the axis EK, KF into which LK divides it are said to be the abscissæ corresponding to that ordinate.

NOTE.—OVAL. A curve line, the two diameters of which are of unequal length, and is allied in form to the ellipse. An ellipse is that figure which is produced by cutting a cone or cylinder in a direction oblique to its axis, and passing through its sides. An oval may be formed by joining different segments of circles, so that their meeting shall not be perceived, but form a continuous curve line. All ellipses are ovals, but all ovals are not ellipses; for the term oval may be applied to all egg-shaped figures, those which are broader at one end than the other, as well as those whose ends are equally curved.

TO DESCRIBE AN ELLIPSE.

[Drawn for this work by L. W. Truesdell, Tinman, Owego, N. Y.]

Original.   •

Fig. 4.

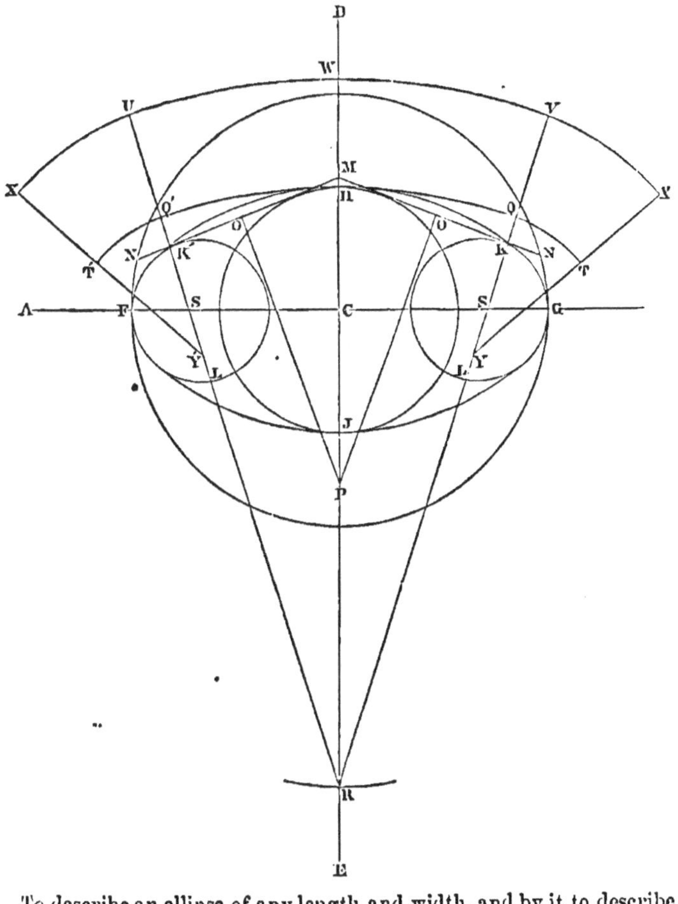

. To describe an ellipse of any length and width, and by it to describe a pattern for the sides of a vessel of any flare.

First draw an indefinite line DE perpendicular to the line AB, and from C, the point of intersection, as a centre, describe a circle FG, having the diameter equal to the length of the ellipse; from the same centre C describe a circle IIJ equal to the width ; then describe the end circles LK′ aud LK, as much less than the width as the width is less than the length ; then draw the lines MN aud MN tangent to the circles K′L, IIJ and KL ; from the middle of the line MN at O erect a perpendicular produced until it intersects the indefinite line DE : from the point of intersection P as a centre, describe the arc K′HK, and with the same sweep of the dividers mark the point R on the line DE ; from the point R draw the lines RU and RV through the points K′ and K where the arc K′HK touches the end circles K′L and KL ; then place one foot of the dividers on the point R and span them to the point II, and describe the arc Q′IIQ, which will be equal in length to the arc K′IIK ; from the same centre R describe the arc UWV the width of the pattern ; then span the dividers the diameter of the end circle KL ; place one foot of the dividers on the line RV, at point Q, and the other at Y as a centre, describe the arc QT the length of the curve line KG, and with the same sweep of the dividers describe the arc T′Q′ from the centre Y′ on the line RU ; then span the dividers from Y′ to U, and from Y′ as a centre, describe the arc UX, and from Y as a centre, describe the arc VX, which completes the description of the pattern.

The more flare you wish the pattern to have, the nearer the centre point R must be to II ; and the less flare, the further the centre point R must be from II ; in the same proportion as you move the centre R towards, or from II, you must move the centre Y towards, or from Q, or which would be the same as spanning the dividers less, or greater, than the diameter of the end circle KL.

TO FIND THE CIRCUMFERENCE OF AN ELLIPSE.

RULE.—Multiply half the sum of the two diameters by 3·1416, and the product will be the circumference.

Example.—Suppose the longer diameter 6 inches and the shorter diameter 4 inches, then 6 added to 4 equal 10, divided by 2 equal 5, multiplied by 3·1416 equal 15·7080 inches circumference.

TO FIND THE AREA OF AN ELLIPSE.

RULE.—Multiply the longer diameter by the shorter diameter, and by ·7854, and the product will be the area.

Example.—Required the area of an ellipse whose longer diameter is 6 inches and shorter diameter 4 inches?

$$6 \times 4 \times ·7854 = 18·8496, \text{ the area.}$$

TO DESCRIBE A RIGHT ANGLED ELBOW.

[Drawn for this work by L. W. Truesdell, Tinman, Owego, N. Y.]

Original.

Fig. 5.

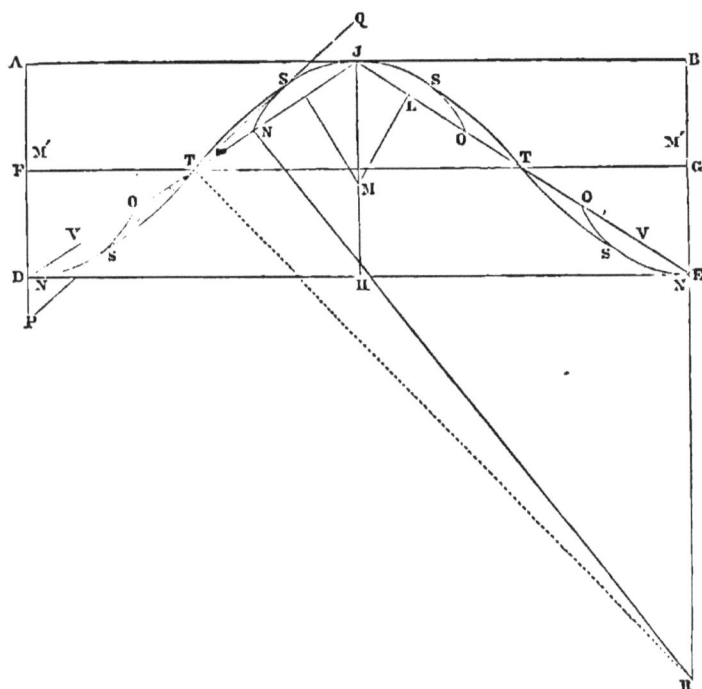

First construct a rectangle ADEB equal in width to the diameter of the elbow, and the length equal to the circumference; then from the point J, the middle of the line AB, draw the line JH, and from the point F, the middle of the line AD, draw the line FG ; from the point J draw two diagonal lines JD and JE ; then span the dividers so as to divide one of these diagonal lines into six equal parts, viz. J, L, O, T, O, V, E ; from the point L erect a perpendicular, produced to the line JH ; from the point of contact M, as a centre, describe the arc NJO for the top of the elbow, and from the points

M' and M' as centres, with the same sweep of the dividers, describe the arcs NO and NO ; then draw an indefinite straight line PQ tangent to the arcs NO and NJ, having the points of contact at S and S ; on this tangent line erect a perpendicular passing through the point N produced until it intersects the line-BE produced ; then place one foot of the dividers on the point of intersection R and span them over the dotted line to the point T, and with the dividers thus spanned describe the arcs TS, TS, TS, and TS ; these arcs and the arcs NO, NJO, and ON will be the right angled elbow required.

TO DESCRIBE A STRAIGHT ELBOW.

[Old Method.]

FIG. 6.

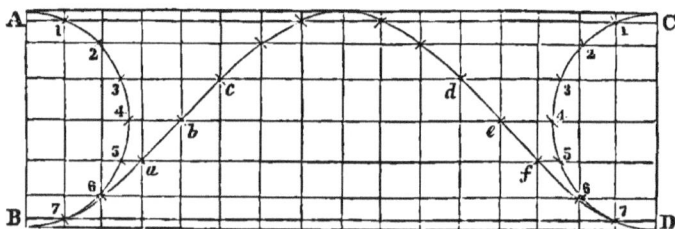

Mark out the length and depth of the elbow, ABCD ; draw a semicircle at each end, as from AB and CD ; divide each semicircle into eight parts ; draw horizontal lines as shown from 1 to 1, 2 to 2, &c. ; divide the circumference or length, ACBD, into sixteen equal parts, and draw perpendicular lines as in figure ; draw a line from *a* to *b* and from *b* to *c*, and on the opposite side from *d* to *e* and *e* to *f;* for the top sweep set the dividers on *fourth* line from top and sweep *two* of the spaces ; the same at the corner ; on space for the remaining sweeps set the dividers so to intersect in the three corners of the spaces marked ✕. The seams must be added to drawing.

TO DESCRIBE A CURVED ELBOW.

[Drawn for this work by L. W. Truesdell, Tinman, Owego, N. Y.]

Original.

Fig. 7.

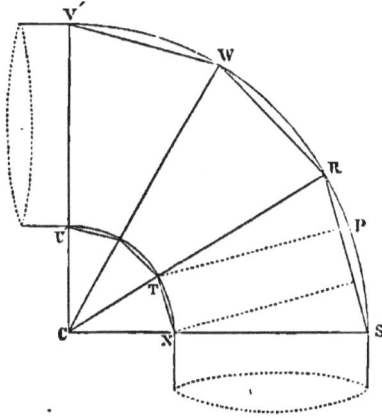

Fig. 8.

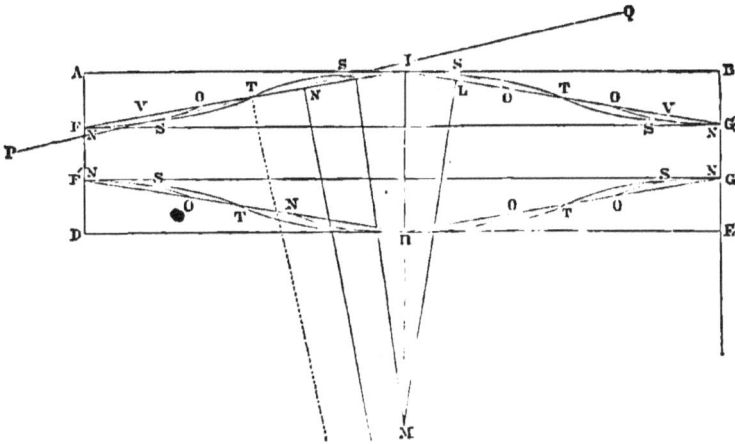

Describe two circles UX and V'S, the curves desired for the elbow, having the distance from U to V' equal to the diameter ; then divide the circle V', W, R and S, into as many sections as desired ; then construct a rectangle, *Fig.* 8, ADEB, the width equal to the width of one section V'W, *Fig.* 7, and the length equal to the circumference of the elbow ; then span the dividers from the point R to the point P at the dotted line, *Fig.* 7, and with the dividers thus spanned mark the points FF' *Fig.* 8, from points A and D, and draw the lines FG and F'G' ; from point I draw the two diagonal lines IF and IG, span the dividers so as to divide one of these diagonal lines into six equal parts, viz. I, L, O, T, O, V, G ; from the point L erect a perpendicular line produced until it intersects the line III produced ; from the point of intersection M, as a centre, describe the arc NIO for the top of the elbow ; with the same sweep of the dividers describe the arcs NO and NO ; then draw an indefinite straight line PQ tangent to the arcs NO and NI, having the points of contact at S and S ; on this tangent line erect a perpendicular line passing through the point N (same as in *Fig.* 5), produced until it intersects the line BE produced ; then place one foot of the dividers on the point of intersection and span them over the dotted line to the point T, (same as in *Fig.* 5), and with the dividers spanned describe the arcs TS, TS, TS, and TS ; these arcs and the arcs NO, NIO and ON, will be one side of the section, and by the same rule the other side of the section may be described at the same time, which will be a pattern to cut the other sections by.

SOLDERING.

For Lead the solder is 1 part tin, 1 to 2 of lead; — for *Tin* 1 to 2 parts tin to 1 of lead ; — for *Zinc* 1 part tin to 1 to 2 of lead ; — for *Pewter* 1 part tin to 1 of lead, and 1 to 2 parts of bismuth.

The surfaces to be joined are made perfectly clean and smooth, and then covered with sal-ammoniac, or resin, or both ; the solder is then applied, being melted in, and smoothed over by the soldering iron.

To Joint Lead Plates.—The joints of lead plates for some purposes are made as follows : — The edges are brought together, hammered down into a sort of channel cut out of wood, and secured with a few tacks. The hollow is then scraped clean with a scraper, rubbed over with candle grease, and a stream of hot lead is poured into it, the surface being afterwards smoothed with a red-hot plumber's iron.

TO DESCRIBE A STRAIGHT ELBOW.

[Another Method for describing a Straight Elbow.]

FIGS. 9 & 10.

Fig. 10.   *Fig.* 9.

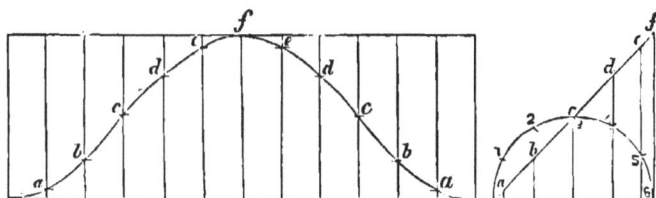

FIG. 9.—Draw a profile of half of the elbow wanted, and mark a semicircle on the line representing the diameter, divide the semicircle into six equal parts, draw perpendicular lines from each division on the circle to the angle line as on figure.

FIG. 10.   Draw the circumference and depth of elbow wanted, and divide into twelve equal parts, mark the height of perpendicular lines of *Fig.* 9 on *Fig.* 10 *a b c* &c.; set your dividers the same as for the semicircle and sweep from *e* to *e* intersecting with *f* and the same from *a* to the *corner*, then set the dividers one-third the circumference and sweep from *e* to *d each side*, and from *a* to *b each side* at bottom; then set your dividers three-fourths of the circumference and sweep from *c* to *d each side* on top, and from *c* to *b* at bottom, and you obtain a more correct pattern than is generally used. Allow for the lap or seam outside of your drawing, and lay out the elbow deep enough to put together by swedge or machine. Be careful in dividing and marking out, and the large end will be true without trimming. The seams must be added to drawing.

To Joint Lead Pipes.—Widen out the end of one pipe with a taper wood drift, and scrape it clean inside; scrape the end of the other pipe outside a little tapered, and insert it in the former: then solder it with common lead solder as before described; or if required to be strong, rub a little tallow over, and cover the joint with a ball of melted lead, holding a cloth (2 or 3 plies of greased bed-tick) on the under side; and smoothing over with it and the plumber's iron.

TO DESCRIBE BEVEL COVERS FOR VESSELS, OR BREASTS FOR CANS.

[Drawn for this work by L. W. TRUESDELL, Tinman, Owego, N. Y.]

FIG. 11.

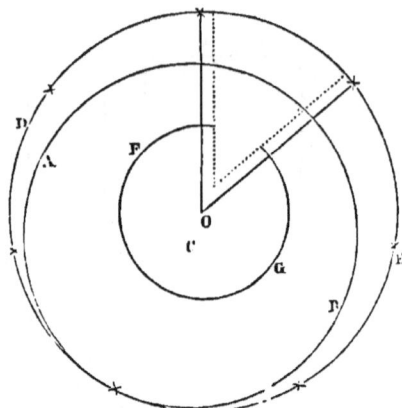

From O as a centre, describe a circle DE larger than the vessel; and from C as a centre, describe a circle AB the size of the vessel, then with the dividers the same as you described the circle the size of the vessel, apply them six times on the circumference of the circle larger than the vessel; for can-breasts describe the circle FG the size you wish for the opening of the breast.

TO DESCRIBE PITCHED COVERS FOR PAILS, &c.

FIG. 12.

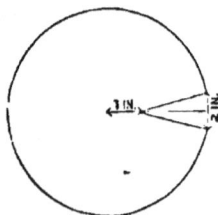

To cut for pitched covers, draw a circle one inch larger than the hoop is in diameter after burring, then draw a line from the centre to

3

the circumference as in the figure, and one inch from the centre and connecting with this line draw two more lines the ends of which shall be one inch on either side of the line first drawn, and then cut out the piece.

TO DESCRIBE AN OVAL BOILER COVER.

[Drawn for this work by L. W. TRUESDELL, Tinman, Owego, N. Y.]

FIG. 13.

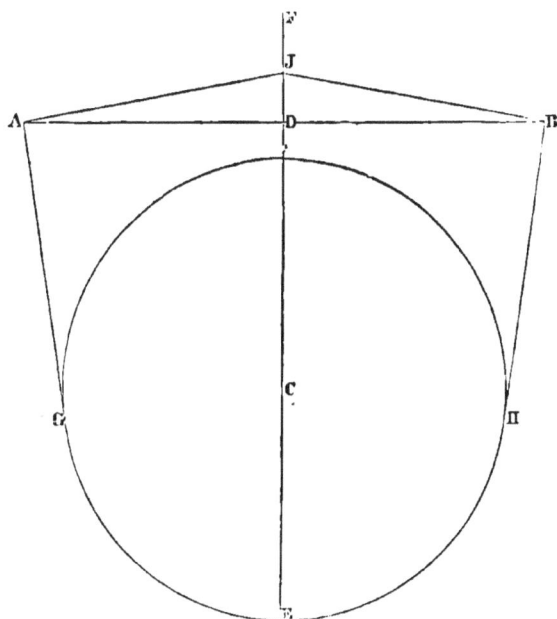

From C as a centre, describe a circle whose diameter will be equal to the width of the boiler outside of the wire, and draw the line AB perpendicular to the line EF, having it pass through the point D, which is one-half of the length of the boiler ; then mark the point J one quarter of an inch or more as you wish, for the pitch of the cover, and apply the corner of the square on the line AB, allowing the blade to fall on the circle at II, and the tongue at the point J ; then draw the lines IIB, BJ, GA and AJ, which completes the description.

TO DESCRIBE A LIP TO A MEASURE.

[Drawn for this work by L. W. Truesdell, Tinman, Owego, N. Y.]

Original.

Fig. 14.

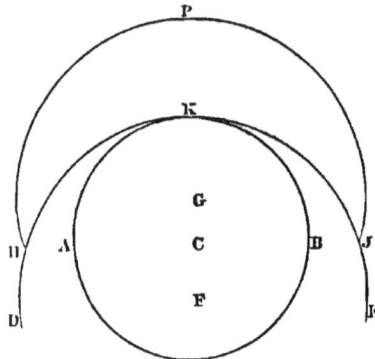

Let the circle AB represent the size of the measure ; span the dividers from K to F three-quarters of the diameter ; describe the semicircle DKE ; move the dividers to G the width of the lip required, and describe the semicircle KPJ, which will be the lip sought.

THE CIRCLE AND ITS SECTIONS.

1. The *Areas of Circles* are to each other as the squares of their diameters ; any circle twice the diameter of another contains four times the area of the other.

2. The *Radius* of a circle is a straight line drawn from the centre to the circumference.

3. The *Diameter* of a circle is a straight line drawn through the centre, and terminated both ways at the circumference.

4. A *Chord* is a straight line joining any two points of the circumference.

5. An *Arc* is any part of the circumference.

6. A *Semicircle* is half the circumference cut off by a diameter.

7. A *Segment* is any portion of a circle cut off by a chord.

8. A *Sector* is a part of a circle cut off by two radii.

TO DESCRIBE A FLARING VESSEL PATTERN, A SET OF PATTERNS FOR A PYRAMID CAKE, OR AN ENVELOPE FOR A CONE.

[Drawn for this work by L. W. TRUESDELL, Tinman, Owego, N. Y.]

Original.

FIG. 15.

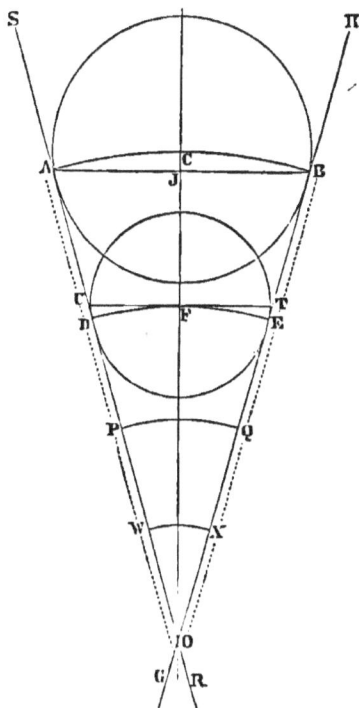

From a point C as a centre, describe a circle AB equal to the large circumference ; with the point F as a centre, the depth of the vessel, describe a circle DE equal to the small circumference ; then draw the lines GH and RS tangent to the circles AB and DE ; from the point of intersection O as a centre, describe the arcs ACB and DFE ; then ADEB will be the size of the vessel, and three such pieces will be an envelope for it, and AJBTFU the altitude ; then by dividing the sector

SOH into sections AB, DE, PQ, and WX, you will have a set of patterns for a pyramid cake; and the sector AOB will be one-third of an envelope for a cone.

In allowing for locks, you must draw the lines parallel to the radii, as represented in the diagram by dotted lines, which will bring the vessel true across the top and bottom.

TO DESCRIBE A CONE OR FRUSTUM.

FIG. 16.

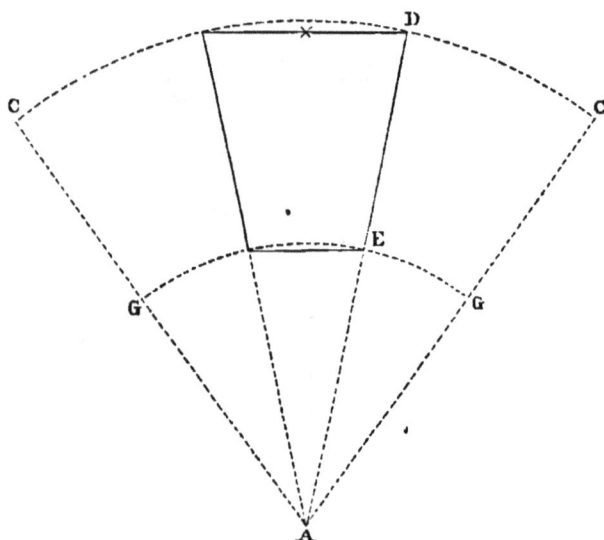

First draw a side elevation of the desired vessel, DE, then from A as a centre describe the arcs CDC and GEG ; after finding the diameter of the top or large end, turn to the table of Diameters and Circumferences, where you will find the true circumference, which you will proceed to lay out on the upper or larger arc CDC, making due allowance for the locks, wire and burr. This is for one piece ; if for two pieces you will lay out only one-half the circumference on the plate ; if for three pieces one-third ; if for four pieces one-fourth ; and so on for any number, remembering to make the allowance for locks, wire and burr on the piece you use for a pattern.

3*

TO DESCRIBE A HEART.

[Drawn for this work by L. W. TRUESDELL, Tinman, Owego, N. Y.]

FIG. 17.

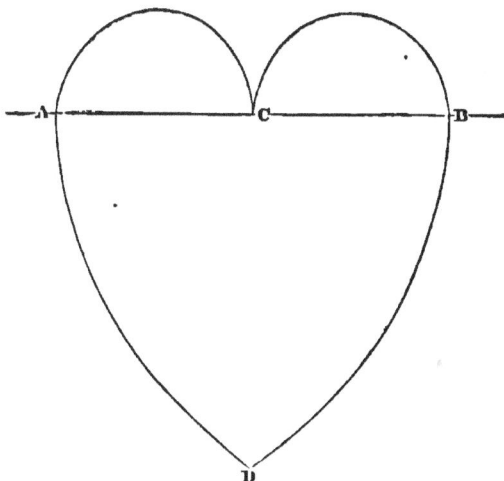

Draw an indefinite line AB ; then span the dividers one-fourth the width you wish the heart, and describe two semicircumferences AC and CB ; span the dividers from A to B, the width of the heart, and describe the lines AD and BD, which completes the description.

CYCLOID.

FIG. 18.

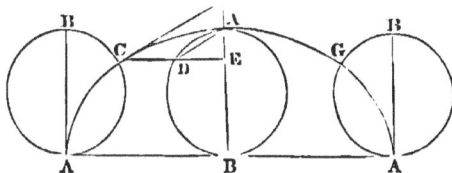

Cycloid, a curve much used in mechanics. It is thus formed :—

If the circumference of a circle be rolled on a right line, beginning at any point A, and continued till the same point A arrive at the line again, making just one revolution, and thereby measuring out a straight line ABA equal to the circumference of a circle, while the point A in the circumference traces out a curve line ACAGA : then this curve is called a cycloid ; and some of its properties are contained in the following lemma.

If the generating or revolving circle be placed in the middle of the cycloid, its diameter coinciding with the axis AB, and from any point there be drawn the tangent CF, the ordinate CDE perpendicular to the axis, and the chord of the circle AD ; then the chief properties are these :

The right line   CD equal to the circular arc   AD ;

The cycloidal arc AC equal to double the chord AD ;

The semi-cycloid ACA equal to double the diameter AB, and

The tangent CF is parallel to the chord AD.

This curve is the line of swiftest descent, and that best suited for the path of the ball of a pendulum.

TO STRIKE THE SIDE OF A FLARING VESSEL.

Fig. 19.

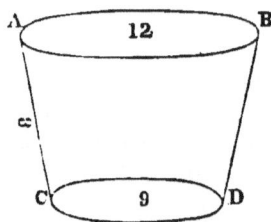

To find the radius of a circle for striking the side of a flaring vessel having the diameters and depth of side given.

RULE. — As the difference between the large and small diameter is to the depth of the side, so is the small diameter to the radius of the circle by which it is struck.

Example. — Suppose ABCD to be the desired vessel, with a top diameter of 12 inches, bottom diameter 9 inches, depth of side 8 inches. Then as $12 - 9 = 3 : 8 : : 9$ to the radius.

$$8 \times 9 = 72 \div 3 = 24 \text{ inches, answer.}$$

TINNING IRON.

Cleanse the metal to be tinned, and rub with a coarse cloth, previously dipped in hydrochloric acid, (muriatic acid) and then rub on French putty with the same cloth. French putty is made by mixing tin filings with mercury.

TO DESCRIBE BEVEL COVERS FOR VESSELS, OR BREASTS FOR CANS.

Fig. 20.

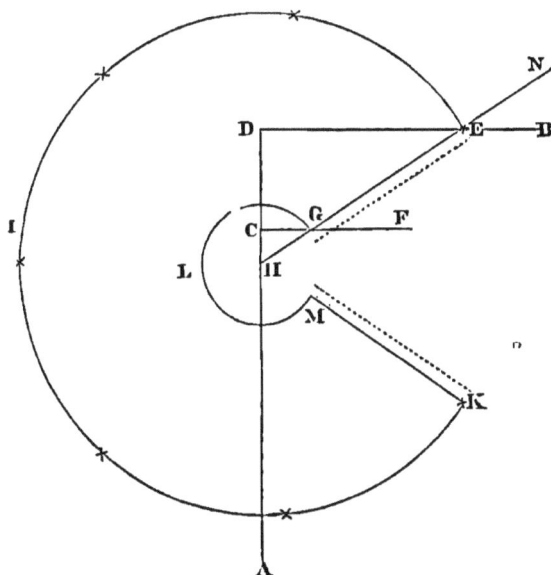

Construct a right angle ADB, and from the point C, the altitude height you wish the breast, erect a perpendicular line F ; then on the line B, mark the point E one-half the diameter of the can ; and on the line F, mark the point G one-half the diameter of the opening in the top of breast ; draw a line N to pass through the points E and G produced until it intersects the line A ; place one foot of the dividers at the point of intersection II, and place the other on the point E, and describe the circle EIK ; span the dividers from the point II to point G, and describe the circle GLM ; then span the dividers from the point D to E, and step them six times on the circle EIK, which gives the size of the breast. Remember to mark the lines for the locks parallel with the radii.

A GOOD SOLDER.

Take 1 lb. of pure Banca tin, and melt it, then add half a pound of clean lead, and when it is melted, stir the mixture gently with a stick or poker, and pour it out into solder strips.

TO FIND THE CENTRE OF A CIRCLE FROM A PART OF THE CIRCUMFERENCE.

[Drawn for this work by L. W. TRUESDELL, Tinman, Owego, N. Y.]

Original.

FIG. 21.

Span the dividers any distance you wish, and place one foot on the circumference AB, and describe the semicircumferences CD, EF, GH, and IK, and through the points of their intersection PQ and RS, draw two indefinite lines LM and NO ; the point of their intersection T, will be the centre desired.

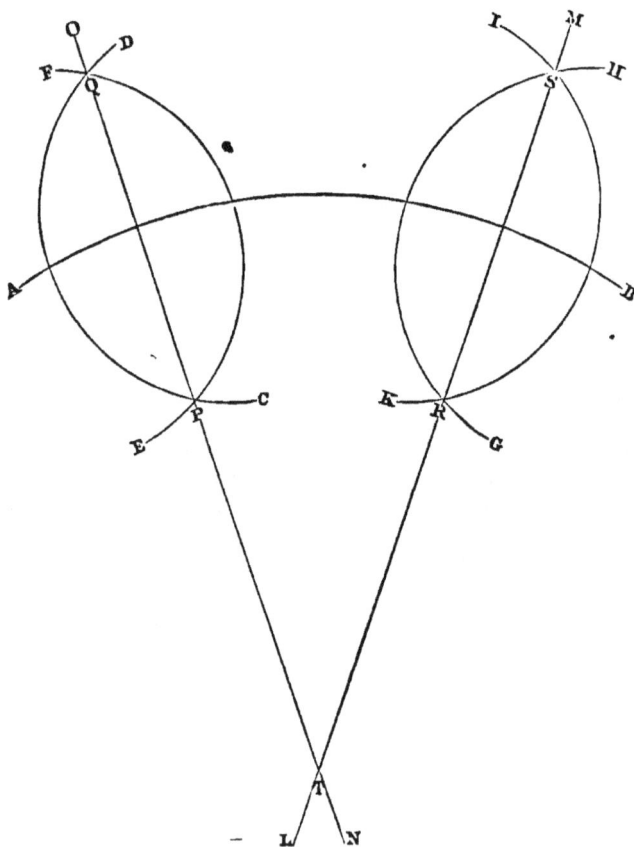

SECTOR, FOR OBTAINING ANGLES.

FIG. 22.

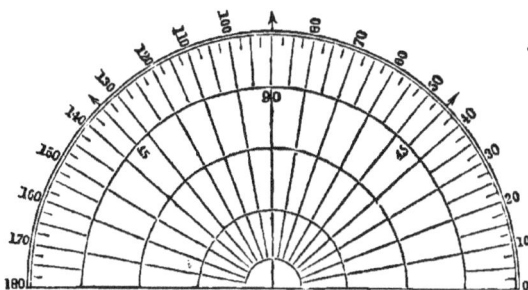

SECTOR, a portion of a circle comprehended between any two radii and their intercepted arcs.— *Similar Sectors* are those whose radii include equal angles.

To find the area of a sector. Say as 360° is to the degrees, &c., in the arc of the sector, so is the area of the whole circle to the area of the sector. Or multiply the radius by the length of the arc, and half the product will be the area.

. TO CONSTRUCT THE FRUSTUM OF A CONE.

Form of flat Plate by which to construct any Frustum of a Cone.

FIG. 23.

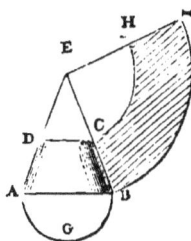

Let ABCD represent the required frustum ; continue the lines AD and BC until they meet at E ; then from E as centre, with the radius EC, describe the arc CH ; also from E, with the radius EB, describe the arc BI ; make BI equal in length to twice AGB, draw the line EI, and BCHI is the form of the plate as required.

RULE FOR STRIKING OUT A CONE OR FRUSTUM.

FIG. 24.

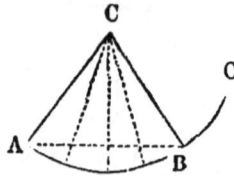

In a conical surface, there may be economy, sometimes, in having the slant height 6 times the radius of base. For a Circle may be wholly cut into conical surfaces, if the angle is 60°, 30°, 15°, &c.

But there is a greater simplicity in cutting it, when the angle is 60°. For instance, take AC equal to the slant height, describe an indefinite arc AO ; with the same opening of the dividers measure from A to B ; draw BC and we have the required sector. This would make the angle C equal 60°. This angle may be divided into two or four equal parts, and we should thus have sectors whose angle would be 30° or 15°, which would not make the vessel very flaring. The accompanying figure gives about the shape of the flar-

FIG. 25.

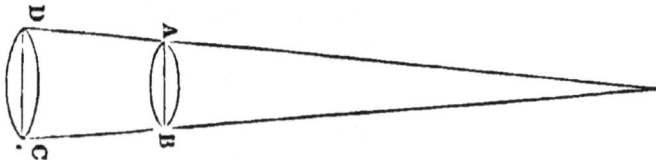

ing vessel when the angle of the sector is 30°.

TO FIND THE CONTENTS OF A PYRAMID OR CONE.

RULE.—Multiply the diameter of the base by itself, and this product by the height, then take one-third of this product for the contents ; to obtain gallons, divide the last result by 231.

EXAMPLE.—Required the cubic inches of a Cone whose base is 8 inches diameter, and height 18 inches.

$8 \times 8 = 64 \times 18 = 1152 \div 3 = 384$ cubic inches, $\div 231 = 1$ gall. $2\frac{3}{4}$ quarts.

HIPPED ROOFS, MILL HOPPERS, &c.

To find the various Angles and proper Dimensions of Materials whereby to construct any figure whose form is the Frustum of a proper or inverted Pyramid, as Hipped Roofs, Mill Hoppers, &c.

FIG. 26.

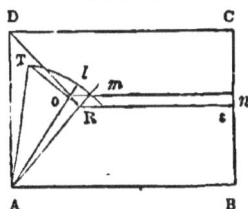

Let ABCD be the given dimensions of plan for a roof, the height RT, also being given ; draw the diagonal AR, meeting the top or ridge Rs on plan ; from R, at right angles with AR and equal to the required height, draw the line RT, then TA, equal the length of the struts or corners of the roof; from A, with the distance AT, describe an arc Tl, continue the diagonal AR until it cuts the arc Tl, through which, and parallel with the ridge Rs, draw the line m n, which determines the required breadth for each side of the roof: from A, meeting the line m n, draw the line Ao, or proper angle for the end of each board by which the roof might require to be covered ; and the angle at T is what the boards require to be made in the direction of their thickness, when the corners or angles require to be mitred.

CONTENTS IN GALLONS OF THE FRUSTUM OF A CONE.

FIGS. 27, 28, 29.

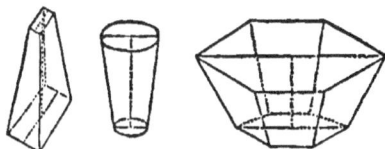

To find the Contents in Gallons of a Vessel, whose diameter is larger at one end than the other, such as a Bowl, Pail, Firkin, Tub, Coffee-pot, &c.

RULE.—Multiply the larger diameter by the smaller, and to the product add one-third of the square of their difference, multiply by the height, and multiply that product by .0034 for Wine Gallons, and by .002785 for Beer.

EXAMPLE.—Required the contents of a Coffee-pot 6 inches diameter at the top, 9 inches at the bottom, and 18 inches high.

large diameter	9	brought up	1026
small do.	6		.0034

	54		4104
⅓ of the square	3		3078

	57		3.4884 Wine Gallons,
height	18		or nearly 3½ gallons.

	456
	57

Carried up   1026

1026 multiplied by .002785 equal 2.8574 *Beer Gallons.*

RULE TO FIND THE CONTENTS IN GALLONS OF ANY SQUARE VESSEL.

RULE.—Take the dimensions in inches and decimal parts of an inch, multiply the length, breadth, and height together, and then multiply the product by .004329 for Wine Gallons, and by .003546 for Ale Gallons.

EXAMPLE.—How many Wine Gallons will a box contain that is 10 feet long, 5 feet wide, and 4 feet deep.

Length in inches,	120	brought up	345600
Breadth in do.	60		.004329

	7200		3110400
Height in inches,	48		691200

	57600		1036800
	28800		1382400

Carried up,	345600	1496.102400 gallons.

or 1496 galls. and 8¼ gills

4

CONTENTS IN GALLONS OF CYLINDRICAL VESSELS.

RULE.—Take the dimensions, in inches and decimal parts of an inch. Square the diameter, multiply it by the length in inches, and then multiply the product by .0034 for Wine Gallons, or by .002785 for Ale Gallons.

EXAMPLE.—How many U. S. Gallons will a Cylindrical Vessel contain, whose diameter is 9 inches, and length 9½ inches?

Diameter,	9	brought up	769.5
	9		.0034
Square Diam.	81		30780
Length,	9.5		23085
	405		2.61630
	729	or 2 gallons and 5 pints.
Carried up,	769.5		

TO ASCERTAIN THE WEIGHTS OF PIPES OF VARIOUS METALS, AND ANY DIAMETER REQUIRED.

Thickness in parts of an inch.	Wrought iron.	Copper.		Lead.		
1-32	·326	11¼ lbs. plate	·38	2 lbs. lead		·483
1-16	·653	23½ "	·76	4 "		·967
3-32	·976	35 "	1·14	5½ "		1·45
1-8	1·3	46½ "	1·52	8 "		1·933
5-32	1·627	58 "	1·9	9¼ "		2·417
3-16	1·95	70 "	2·28	11 "		2·9
7-32	2·277	80½ "	2·66	13 "		3·383
1-4	2·6	93 "	3·04	15 "		3.867

Rule.—To the interior diameter of the pipe, in inches, add the thickness of the metal; multiply the sum by the decimal numbers opposite the required thickness and under the metal's name; also by the length of the pipe in feet, and the product is the weight of the pipe in lbs.

1. Required the weight of a copper pipe whose interior diameter is 7½ inches, its length 6¼ feet, and the metal 1-8 of an inch in thickness.

$$7 \cdot 5 + \cdot 125 = 7 \cdot 625 \times 1 \cdot 52 \times 6 \cdot 25 = 72 \cdot 4 \text{ lbs.}$$

2. What is the weight of a leaden pipe 18½ feet in length, 3 inches interior diameter, and the metal ¼ of an inch in thickness?

$$3 + \cdot 25 = 3 \cdot 25 \times 3 \cdot 867 \times 18 \cdot 5 = 232 \cdot 5 \text{ lbs.}$$

TIN PLATES.

Size, Length, Breadth, and Weight.

BRAND MARK.	No. of Sheets in Box.	Length and Breadth.	Weight per Box.	
		Inches.Inches.	Cwt. qr. lbs.	
1 C	225	14 by 10	1 0 0	
1 x	225	14 by 10	1 1 0	
1 xx	225	14 by 10	1 1 21	Each 1x advances
1 xxx	225	14 by 10	1 2 14	$1.75 to $2.00
1 xxxx	225	14 by 10	1 3 7	
1 xxxxx	225	14 by 10	2 0 0	
1 xxxxxx	225	14 by 10	2 0 21	
D C	100	17 by 12½	0 3 14	
D x	100	17 by 12½	1 0 14	
D xx	100	17 by 12½	1 1 7	
D xxx	100	17. by 12½	1 2 0	
D xxxx	100	17 by 12½	1 2 21	
D xxxxx	100	17 by 12½	1 3 14	
D xxxxxx	100	17 by 12½	2 0 7	
S D C	200	15 by 11	1 1 27	
S D x	200	15 by 11	1 2 20	
S D xx	200	15 by 11	1 3 13	
S D xxx	200	15 by 11	2 0 6	
S D xxxx	200	15 by 11	2 0 27	
S D xxxxx	200	15 by 11	2 1 20	
S D xxxxxx	200	15 by 11	2 2 13	
			about	
T T Taggers,	225	14 by 10	1 0 0	
1 C	225	12 by 12		
1 x	225	12 by 12		
1 xx	225	12 by 12		
1 xxx	225	12 by 12		
1 xxxx	225	12 by 12		
1 C	112	14 by 20		About the same weight per Box, as the plates above of similar brand, 14 by 10.
1 x	112	14 by 20		
1 xx	112	14 by 20		
1 xxx	112	14 by 20		
1 xxxx	112	14 by 20		
Leaded or 1 C	112	14 by 20	1 0 0	For Roofing.
Ternes 1 x	112	14 by 20	1 1 0	

(Note in right margin spanning D, S D, and Taggers groups:) In addition, a great variety of sizes are imported for special purposes, usually costing a little more in proportion than those which are esteemed regular sizes.

OIL CANISTERS, *(from 2½ to 125 galls.)* WITH THE QUANTITY AND QUALITY OF TIN REQUIRED FOR CUSTOM WORK.

Galls.	Quantity and Quality.		Galls.	Quantity and Quality.
2½	2 Plates, I X in body.		33	13½ Plates, IX in body, 3 breadths high.
3½	2 " S DX "			
5½	2 " DX "		45	13½ Plates, S D X in body.
8	4 " IX "		60	13½ " D X "
10	3½ " DX "		90	15¼ " D X " *
15	4 . " DX "		125	20 " D X "

* The bottom tier of plates to be placed lengthwise.

WEIGHT OF WATER.

1	cubic inch	is equal to	.03617	pounds.
12	cubic inches	is equal to	.434	pounds.
1	cubic foot	is equal to	62.5	pounds.
1	cubic foot	is equal to	7.50	U. S. gallons.
1.8	cubic feet	is equal to	112.00	pounds.
35.84	cubic feet	is equal to	2240.00	pounds.
1	Cylindrical inch ..	is equal to	.02842	pounds.
12	Cylindrical inches.	is equal to	.341	pounds.
1	Cylindrical foot ..	is equal to	49.10	pounds.
1	Cylindrical foot ..	is equal to	6.00	U. S. Gallons.
2.282	Cylindrical feet ..	is equal to	112.00	pounds.
45.64	Cylindrical feet ..	is equal to	2240.00	pounds.
11.2	Imperial gallons ..	is equal to	112.00	pounds.
224	Imperial gallons ..	is equal to	2240.00	pounds.
13.44	United States galls.	is equal to	112.00	pounds.
268.8	United States galls.	is equal to	2240.00	pounds.

Centre of pressure is at two-thirds depth from surface.

DECIMAL EQUIVALENTS TO THE FRACTIONAL PARTS OF A GALLON, OR AN INCH.

[The Inch, or Gallon, being divided into 32 parts.]

[In multiplying decimals it is usual to drop all but the two or three first figures.]

Decimals.	Gallon. or Inch.	Gills.	Pints.	Quarts.	Decimals.	Gallon. or Inch.	Gills.	Pints.	Quarts.	Decimals.	Gallon. or Inch.	Gills.	Pints.	Quarts.
.03125	1-32	1	1/4	1/8	.375	3-8	12	3	1 1/2	.71875	23-32	23	5 3/4	2 7/8
.0625	1-16	2	1/2	1/4	.40625	13-32	13	3 1/4	1 5/8	.75	3-4	24	6	3
.09375	3-32	3	3/4	3/8	.4375	7-16	14	3 1/2	1 3/4	.78125	25-32	25	6 1/4	3 1/8
.125	1-8	4	1	1/2	.46875	15-32	15	3 3/4	1 7/8	.8125	13-16	26	6 1/2	3 1/4
.15625	5-32	5	1 1/4	5/8	.5	1-2	16	4	2	.84375	27-32	27	6 3/4	3 3/8
.1875	3-16	6	1 1/2	3/4	.53125	17-32	17	4 1/4	2 1/8	.875	7-8	28	7	3 1/2
.21875	7-32	7	1 3/4	7/8	.5625	9-16	18	4 1/2	2 1/4	.90625	29-32	29	7 1/4	3 5/8
.25	1-4	8	2	1	.59375	19-32	19	4 3/4	2 3/8	.9375	15-16	30	7 1/2	3 3/4
.28125	9-32	9	2 1/4	1 1/8	.625	5-8	20	5	2 1/2	.96875	31-32	31	7 3/4	3 7/8
.3125	5-16	10	2 1/2	1 1/4	.65625	21-32	21	5 1/4	2 5/8	1.000	1	32	8	4
.34375	11-32	11	2 3/4	1 3/8	.6875	11-16	22	5 1/2	2 3/4					

APPLICATION. Required the *gallons* in any Cylindrical Vessel. Suppose a vessel 9 1-2 inches deep, 9 inches diameter, and contents 2·6163, that is, 2 gallons and 61 hundredth parts of a gallon, now to ascertain this decimal of a gallon refer to the above Table, for the decimal that is nearest, which is ·625, opposite to which is 5-8ths of a gallon, or 20 gills, or 5 pints, or 2 1-2 quarts, consequently the vessel contains 2 gallons and 5 pints.

INCHES. To find what part of an inch the decimal ·708 is. Refer to the above Table for the decimal that is nearest, which is ·71875, opposite to which is 23-32, or nearly 3-4ths of an inch.

A TABLE

CONTAINING THE

DIAMETERS, CIRCUMFERENCES, AND AREAS OF CIRCLES,

AND THE

CONTENT OF EACH IN GALLONS AT 1 FOOT IN DEPTH.

~~~~~~~~~

UTILITY OF THE TABLE.

EXAMPLES.

1. Required the circumference of a circle, the diameter being *five* inches ?

In the column of circumferences opposite the given diameter, stands 15·708* inches, the circumference required.

2. Required the capacity, in gallons, of a can the diameter being 6 feet and depth 10 feet?

In the fourth column from the given diameter stands 211.4472* being the content of a can 6 feet in diameter and 1 foot in depth, which being multipled by 10 gives the required content, two thousand one hundred fourteen and a half gallons.

3, Any of the areas in feet multiplied by .03704, the product equal the number of cubic yards at 1 foot in depth.

4. The area of a circle in inches multiplied by the length or thickness in inches, and by .263, the product equal the weight in pounds of cast iron.

---

* See opposite page (page 40) for Decimal Equivalents to the Fractional parts of a Gallon, and an Inch.

4*

DIAMETERS AND CIRCUMFERENCES OF CIRCLES, AND THE CONTENT IN GALLONS AT 1 FOOT IN DEPTH.

[Area in Inches.]

Diam.	Circ. in.	Area. in.	Gallons.	Diam.	Circ. in.	Area. in.	Gallons.
1 in.	3·1416	·7854	·04084	6½	20·420	33·183	1·72552
⅛	3·5343	·9940	·05169	⅝	20·813	34·471	1·79249
¼	3 9270	1·2271	·06380	¾	21·205	35·784	1·86077
⅜	4·3197	1·4848	·07717	⅞	21·598	37·122	1·93034
½	4·7124	1·7671	·09188	7 in.	21·991	38·484	2·00117
⅝	5·1051	2 0739	·10784	⅛	22·383	39·871	2·07329
¾	5·4978	2·4052	·12506	¼	22·776	41·282	2·14666
⅞	5·8905	2·7611	·14357	⅜	23·169	42·718	2·22134
2 in.	6·2832	3·1416	·16333	½	23·562	44·178	2.29726
⅛	6 6759	3·5465	·18439	⅝	23·954	45·663	2·37448
¼	7·0686	3·9760	·20675	¾	24·347	47·173	2·45299
⅜	7 4613	4·4302	·23036	⅞	24·740	48·707	2·53276
½	7·8540	4·9087	·25522	8 in.	25·132	50·265	2·61378
⅝	8·2467	5·4119	·28142	⅛	25·515	51·848	2·69609
¾	8·6394	5·9395	·30883	¼	25·918	53 456	2·77971
⅞	9·0321	6·4918	·33753	⅜	26·310	55·088	2·86458
3 in.	9·4248	7·0686	·36754	½	26·703	56·745	2·95074
⅛	9·8175	7·6699	·39879	⅝	27·096	58 426	3·03815
¼	10·210	8.2957	·43134	¾	27·489	60·132	3·12686
⅜	10·602	8·9462	·46519	⅞	27·881	61·862	3·21682
½	10·995	9·6211	·50029	9 in.	28·274	63 617	3·30808
⅝	11·388	10·320	·53664	⅛	28·667	65·396	3·40059
¾	11·781	11·044	·57429	¼	29·059	67·200	3·49440
⅞	12 173	11·793	·61324	⅜	29·452	69·029	3·58951
4 in.	12·566	12·566	·65343	½	29·845	70·882	3·68586
⅛	12·959	13·364	·69493	⅝	30·237	72·759	3·78347
¼	13·351	14·186	·73767	¾	30 630	74·662	3·88242
⅜	13·744	15·033	·78172	⅞	31·023	76·588	3·98258
½	14·137	15·904	·82701	10 in	31·416	78 540	4·08408
⅝	14·529	16·800	·87360	⅛	31·808	80·515	4·18678
¾	14·922	17·720	·92144	¼	32·201	82·516	4·29083
⅞	15·315	18·665	·97058	⅜	32 594	84·540	4·39608
5 in.	15·708	19·635	1·02102	½	32·986	86·590	4·50268
⅛	16·100	20·629	1·07271	⅝	33·379	88·664	4·61053
¼	16·493	21·647	1·12564	¾	33·772	90 762	4·71962
⅜	16·886	22·690	1·17988	⅞	34·164	92 885	4·82846
½	17·278	23·758	1·23542	11 in.	34 557	95·033	4·94172
⅝	17·671	24·850	1·29220	⅛	34·950	97·205	5·05466
¾	18·064	25·967	1·35028	¼	35 343	99·402	5·16890
⅞	18·457	27·108	1·40962	⅜	35·735	101·623	5 28439
6 in.	18·849	28·274	1·47025	½	36·128	103 869	5·40119
⅛	19·242	29·464	1·53213	⅝	36·521	106·139	5·51923
¼	19·635	30·679	1·59531	¾	36·913	108·434	5·63857
⅜	20·027	31·919	1·65979	⅞	37·306	110·753	5·75916

DIAMETERS AND CIRCUMFERENCES OF CIRCLES, AND THE CONTENT IN GALLONS AT 1 FOOT IN DEPTH.

[*Area in Feet.*]

Diam.	Circ.	Area in ft.	Gallons.	Diam.	Circ.	Area in ft.	Gallons.
Ft. In.	Ft. In.		1 ft. in depth	Ft. In.	Ft. In.		1 ft. in depth
1	3 1½	·7854	5·8735	4 6	14 1½	15·9043	118·9386
1 1	3 4⅝	.9217	6·8928	4 7	14 4⅝	16·4986	123·3830
1 2	3 8	1·0690	7·9944	4 8	14 7⅞	17·1041	127·9112
1 3	3 11	1·2271	9·1766	4 9	14 11	17·7205	132·5209
1 4	4 2¼	1·3962	10·4413	4 10	15 2¼	18·3476	137·2105
1 5	4 5⅝	1·5761	11·7866	4 11	15 5¼	18·9858	142·0582
1 6	4 8½	1·7671	13 2150				
1 7	4 11⅝	1·9689	14 7241	5	15 8½	19·6350	146·8384
1 8	5 2⅝	2·4816	16·3148	5 1	15 11½	20·2947	151·7718
1 9	5 5⅞	2·4052	17·9870	5 2	16 2¾	20·9656	156·7891
1 10	5 9	2 6398	19·7414	5 3	16 5⅝	21·6475	161·8886
1 11	6 2¼	2 8852	21·4830	5 4	16 9	22·3400	167·0674
				5 5	17 0¼	23·0437	172·3300
2	6 3¾	3·1416	23·4940	5 6	17 3¼	23 7583	177·6740
2 1	6 6½	3·4087	25·4916	5 7	17 6¾	24·4835	183·0973
2 2	6 9⅝	3 6869	27 5720	5 8	17 9⅝	25·2199	188·6045
2 3	7 0¾	3·9760	29·7340	5 9	18 0⅝	25·9672	194 1930
2 4	7 3⅞	4·2760	32·6976	5 10	18 3⅞	26·7251	199·8610
2 5	7 7	4·5869	34·3027	5 11	18 7⅞	27·4943	205·6133
2 6	7 10¼	4·9087	36·7092				
2 7	8 1⅜	5·2413	39·1964	6	18 10¼	28·2744	211·4472,
2 8	8 4½	5·5850	41 7668	6 3	19 7½	30·6796	229·4342
2 9	8 7⅝	5·9395	44·4179	6 6	20 4⅞	33·1831	248·1564
2 10	8 10¾	6·3049	47·1505	6 9	21 2⅝	35·7847	267·6122
2 11	9 1⅞	6·6813	49·9654				
				7	21 11⅞	38·4846	287·8032
3	9 5	7·0686	52·8618	7 3	22 9¼	41·2825	308·7270
3 1	9 8¼	7·4666	55·8382	7 6	23 6¾	44·1787	330·3859
3 2	9 11⅜	7 8757	58·8976	7 9	24 4⅛	47·1730	352·7665
3 3	10 2½	8·2957	62 0386				
3 4	10 5⅝	8·7265	65·2602	8	25 1¼	50 2656	375·9062
3 5	10 8¾	9·1683	68 5193	8 3	25 11	53 4562	399·7668
3 6	10 11⅞	9·6211	73·1504	8 6	26 8¾	56·7451	424·3625
3 7	11 3	10·0846	75·4186	8 9	27 5¾	60·1321	449·2118
3 8	11 6¼	10·5591	78·9652				
3 9	11 9⅜	11·0446	82 5959	9	28 3¼	63·6174	475·7563
3 10	12 5½	11·5409	86 3074	9 3	29 0⅝	67·2007	502·5536
3 11	12 3⅝	12·0481	90·1004	9 6	29 10¼	70·8823	530·0861
				9 9	30 7½	74·6620	558·3522
4	12 6¾	12·5664	93 9754				
4 1	12 9⅞	13·0952	97·9310				
4 2	13 1	13·6353	101·9701	10	31 5	78·5400	587·3534
4 3	13 4¼	14·1862	103·0300	10 3	32 2⅝	82·5160	617·0876
4 4	13 7⅜	14·7479	110 2907	10 6	32 11⅝	86·5903	647·5568
4 5	13 10½	15·3206	114·5735	10 9	33 9¼	90·7627	678·2797

Diam.		Circ.		Area in ft.	Gallons.	Diam.		Circ.		Area in ft.	Gallons.
Ft.	In.	Ft.	In.		1 ft. in depth	Ft.	In.	Ft.	In.		1 ft. in depth
11		34	6 5/8	95·0334	710·6977	21		65	11 5/8	346·3614	2590·2290
11	3	35	4 1/8	99·4021	743·3686	21	3	66	9	354·6571	2652·2532
11	6	36	1 1/2	103·8691	776·7746	21	6	67	6 1/2	363·0511	2715·0413
11	9	36	10 7/8	108·4342	810·9143	21	9	68	3 7/8	371·5432	2778·5486
12		37	8 3/8	113·0976	848·1890	22		69	1 3/8	380·1336	2842·7910
12	3	38	5 3/4	117·8590	881·3966	22	3	69	10 3/4	388·8220	2907·7664
12	6	39	3 1/4	122·7187	917·7395	22	6	70	8 1/4	397·6087	2973·4889
12	9	40	0 5/8	127·6765	954·8159	22	9	71	5 5/8	406·4935	3039·9209
13		40	10	132·7326	992·6274	23		72	3	415·4766	3107·1001
13	3	41	7 1/2	137·8867	1031·1719	23	3	73	0 1/2	424·5577	3175·0122
13	6	42	4 7/8	143·1391	1070·4514	23	6	73	9 7/8	433·7371	3243·6595
13	9	43	2 1/4	148·4896	1108·0645	23	9	74	7 1/4	443·0146	3313·0403
14		43	11 3/4	153·9384	1151·2129	24		75	4 3/4	452·3904	3383·1563
14	3	44	9 1/8	159·4852	1192·6940	24	3	76	2 1/8	461·8642	3454·0051
14	6	45	6 5/8	165·1303	1234·9104	24	6	76	11 5/8	471·4363	3525·5929
14	9	46	4	170·8735	1277·8615	24	9	77	9	481·1065	3597·9068
15		47	1 1/2	176·7150	1321·5454	25		78	6 3/8	490·8750	3670·9596
15	3	47	10 7/8	182 6545	1365·9634	25	3	79	3 7/8	500·7415	3744·7452
15	6	48	8 1/4	188·6923	1407·5165	25	6	80	1 1/4	510·7063	3819·2657
15	9	49	5 3/4	194·8282	1457·0032	25	9	80	10 3/4	520·7692	3894·5203
16		50	3 1/8	201·0624	1503·6250	26		81	8 1/8	530·9304	3970·5098
16	3	51	0 1/2	207·3946	1550·9797	26	3	82	5 1/4	541·1896	4047·2322
16	6	51	10	213·8251	1599.0696	26	6	83	3	551·5471	4124·6898
16	9	52	7 3/8	220·3537	1647·8930	26	9	84	0 3/8	562·0027	4202·9610
17		53	4 7/8	226·9806	1697·4516	27		84	9 7/8	572·5566	4281·8072
17	3	54	2 1/8	233·7055	1747·7431	27	3	85	8 1/8	583·2085	4361·4664
17	6	54	11 5/8	240·5287	1798 7698	27	6	86	4 5/8	593·9587	4441·8607
17	9	55	9 1/8	247·4500	1850·5301	27	9	87	2 1/8	604·8070	4522·9886
18		56	6 1/2	254·4696	1903·0254	28		87	11 1/2	615·7536	4604·8517
18	3	57	4	261·5872	1956·2537	28	3	88	9	626·7982	4686·4876
18	6	58	1 3/8	268·8031	2010·2171	28	6	89	6 3/8	637·9411	4770·7787
18	9	58	10 3/4	276·1171	2064·9140	28	9	90	3 3/4	649·1821	4854 8434
19		59	8 1/4	283·5294	2120·3462	29		91	1 1/4	660·5214	4939·6432
19	3	60	5 5/8	291·0397	2176·5113	29	3	91	10 5/8	671·9587	5025·1759
19	6	61	3 1/8	298·6483	2233 2914	29	6	92	8 1/8	683·4943	5111·4487
19	9	62	0 1/2	306·3550	2291·0452	29	9	93	5 1/2	695·1280	5198·4451
20		62	9 7/8	314·1600	2349·4141	30		94	2 7/8	706·8600	5286·1818
20	3	63	7 3/8	322 0630	2408·5159	30	3	95	0 3/8	718·6900	5374·6512
20	6	64	4 3/4	330·0643	2468·3528	30	6	95	9 3/4	730·6183	5463·8558
20	9	65	2 1/4	338·1637	2528 9233	30	9	96	7 1/4	742·6447	5553·7940

CAPACITY OF CANS ONE INCH DEEP.

UTILITY OF THE TABLE.

Required the contents of a vessel, diameter 6 7-10*ths* inches, depth 10 inches?
By the table a vessel 1 inch deep and 6 and 7-10*ths* inches diameter contains
.15 (hundredths) of a gallon, then .15 × 10 = 1.50 or 1 gallon and 2 quarts.
Required the contents of a can, diameter 19 8-10*ths* inches, depth 30 inches?
By the table a vessel 1 inch deep and 19 and 8-10*ths* inches diameter contains
1 gallon and .33 (hundredths), then 1.33 × 30 = 39.90 or nearly 40 gallons.
Required the depth of a can whose diameter is 12 and 2-10*ths* inches, to contain 16 gallons.
By the table a vessel 1 inch deep and 12 and 2-10*ths* inches diameter contains
.50 (hundredths of a gallon), then 16 ÷ .50 = 32 inches the depth required, viz:
.50) 16 (32 × .50 = 16 gallons.

Diameter.		$\frac{1}{10}$	$\frac{2}{10}$	$\frac{3}{10}$	$\frac{4}{10}$	$\frac{5}{10}$	$\frac{6}{10}$	$\frac{7}{10}$	$\frac{8}{10}$	$\frac{9}{10}$
3	.03	.03	.03	.03	.03	.04	.04	.04	.04	.05
4	.05	.05	.05	.05	.06	.06	.07	.07	.07	.08
5	.08	.08	.08	.09	.09	.10	.10	.11	.11	.11
6	.12	.12	.12	.13	.13	.14	.14	.15	.15	.16
7	.16	.17	.17	.18	.18	.19	.19	.20	.20	.21
8	.21	.22	.22	.23	.23	.24	.25	.25	.26	.26
9	.27	.28	.28	.29	.30	.30	.31	.31	.32	.33
10	.34	.34	.35	.36	.36	.37	.38	.38	.39	.40
11	.41	.41	.42	.43	.44	.44	.45	.46	.47	.48
12	.48	.49	.50	.51	.52	.53	.53	.54	.55	.56
13	.57	.58	.59	.60	.60	.61	.62	.63	.64	.65
14	.66	.67	.68	.69	.70	.71	.72	.73	.74	.75
15	.76	.77	.78	.79	.80	.81	.82	.83	.84	.85
16	.87	.88	.89	.90	.91	.92	.93	.94	.95	.97
17	.98	.99	1.005	1.017	1.028	1.040	1.051	1.063	1.075	1.086
18	1.101	1.113	1.125	1.138	1.150	1.162	1.170	1.187	1.200	1.211
19	1.227	1.240	1.253	1 266	1.279	1.292	1.304	1.317	1.330	1.343
20	1.360	1.373	1.385	1.400	1.414	1.428	1.441	1.455	1.478	1.482
21	1.499	1.513	1.527	1.542	1.556	1.570	1.585	1.600	1.612	1.630
22	1.645	1.660	1.675	1.696	1.705	1.720	1.735	1 750	1.770	1.780
23	1.798	1.814	1.830	1.845	1.861	1.876	1.892	1.908	1.923	1.940
24	1.958	1.974	1.991	2.007	2.023	2.040	2.056	2.072	2.096	2.105
25	2.125	2.142	2.159	2.176	2.193	2.210	2 227	2.244	2.261	2.280
26	2.298	2.316	2.333	2.351	2.369	2.386	2.404	2.422	2.440	2.460
27	2.478	2.496	2.515	2.533	2.552	2.570	2.588	2.607	2.625	2.643
28	2.665	2.684	2.703	2.722	2.741	2.764	2.780	2.800	2.820	2.836
29	2.859	2.879	2.898	2.918	2.938	2.958	2.977	2.997	3.017	3.036
30	3.060	3.080	3.100	3.121	3.141	3.162	3.182	3.202	3.223	3.245
31	3.267	3.288	3.309	3.330	3.351	3.372	3.393	3.414	3.436	3.457
32	3.481	3.503	3.524	3.543	3.568	3.590	3.612	3.633	3.655	3.689
33	3.702	3.725	3.747	3.773	3.795	3.814	3.837	3 860	3.882	3.904
34	3.930	3.953	3.976	4.003	4.022	4.046	4.070	4.092	4.115	4.140
35	4.165	4.188	4.212	4.236	4.260	4.284	4 307	4.331	4.355	4.380
36	4.406	4.430	4.455	4.483	4.503	4.528	4.553	4 577	4.602	4.626
37	4.654	4.679	4.704	4.730	4.755	4.780	4.805	4.834	4.855	4.880
38	4.909	4.935	4.961	4.987	5.012	5.038	5.064	5.090	5.120	5.142
39	5.171	5.197	5.224	5.250	5.277	5.304	5 330	5.357	5.383	5.410
40	5.440	5.467	5.491	5.521	5.548	5.576	5.603	5.630	5.657	5.684

CRYSTALLIZED TIN-PLATE.

Crystallized tin-plate, is a variegated primrose appearance, produced upon the surface of tin-plate, by applying to it in a heated state some dilute nitro-muriatic acid for a few seconds, then washing it with water, drying, and coating it with lacker. The figures are more or less beautiful and diversified, according to the degree of heat, and relative dilution of the acid. Place the tin-plate, slightly heated, over a tub of water, and rub its surface with a sponge dipped in a liquor composed of four parts of aquafortis, and two of distilled water, holding one part of common salt or sal ammoniac in solution. Whenever the crystalline spangles seem to be thoroughly brought out, the plate must be immersed in water, washed either with a feather or a little cotton (taking care not to rub off the film of tin that forms the feathering), forthwith dried with a low heat, and coated with a lacker varnish, otherwise it loses its lustre in the air. If the whole surface is not plunged at once in cold water, but if it be partially cooled by sprinkling water on it, the crystallization will be finely variegated with large and small figures. Similar results will be obtained by blowing cold air through a pipe on the tinned surface, while it is just passing from the fused to the solid state.

TINNING.

1. Plates or vessels of brass or copper, boiled with a solution of stannate of potassa, mixed with turnings of tin, become, in the course of a few minutes, covered with a firmly attached layer of pure tin.—2. A similar effect is produced by boiling the articles with tin filings and caustic alkali, or cream of tartar. In the above way, chemical vessels made of copper or brass may be easily and perfectly tinned.

NEW TINNING PROCESS.

The articles to be tinned are first covered with dilute sulphuric acid, and when quite clean are placed in warm water, then dipped in a solution of muriatic acid, copper and zinc, and then plunged into a tin bath to which a small quantity of zinc has been added. When the tinning is finished, the articles are taken out and plunged into boiling water. The operation is completed by placing them in a very warm sand bath. This last process softens the iron.

KUSTITIEN'S METAL FOR TINNING.

Malleable iron 1 pound, heat to whiteness ; add 5 ounces regulus of antimony, and Molucca tin 24 pounds.

RECEIPTS

FOR THE USE OF

JAPANNERS, VARNISHERS,

.

BUILDERS AND MECHANICS,

AND FOR

OTHER USEFUL AND IMPORTANT PURPOSES

IN THE

PRACTICAL ARTS.

PRACTICAL RECEIPTS.

[The following Receipts are selected from " Ure's Dictionary," " Cooley's Cyclopedia," " Muspratt's Chemistry," and other valuable sources.]

JAPANNING AND VARNISHING.

Japanning is the art of covering bodies by grounds of opaque colors in varnish, which may be afterwards decorated by printing or gilding, or left in a plain state. It is also to be looked upon in another sense, as that of ornamenting coaches, snuff boxes, screens, &c. All surfaces to be japanned must be perfectly clean, and leather should be stretched on frames. Paper should be stiff for japanning.

The French prime all their japanned articles, the English do not. This priming is generally of common size. Those articles that are primed thus, never endure as well as those that receive the japan coating on the first operation, and thus it is that those articles of japan work that are primed with size when they are used for some time, crack, and the coats of japan fly off in flakes.

A solution of strong isinglass size and honey, or sugar candy, makes a good japan varnish to cover water colors on gold grounds.

A pure white priming for japanning, for the cheap method, is made with parchment size, and one-third of isinglass, laid on very thin and smooth. It is the better for three coats, and when the last coat is dry, it is prepared to receive the painting or figures. Previous to the last coat, however, the work should be smoothly polished. When wood or leather is to be japanned, and no priming used, the best plan is to lay on two or three coats of varnish made of seed-lac and resin, two ounces each, dissolved in alcohol and strained through a cloth. This varnish should be put on in a warm place, and the work to be varnished should, if possible, be warm also, and all dampness should be avoided, to prevent the varnish from being chilled. When the work is prepared with the above composition and dry, it is fit for the proper japan to be laid on. If the ground is not to be white the best varnish now to be used is made of shellac, as it is the best vehicle for all kind of colors. This is made in the proportions of the best shellac, five ounces, made into powder, steeped in a quart of alcohol, and kept at a gentle heat for two or three days and shaken frequently, after which the solution

5

must be filtered through a flannel bag, and kept in a well corked bottle for use. This varnish for hard japanning on *copper* or *tin* will stand for ever, unless fire or hammer be used to burn or beetle it off.

The color to be used with shellac varnish may be of any pigments whatever to give the desired shade, as this varnish will mix with any color.

WHITE JAPAN GROUNDS.

To form a hard, perfect white ground is no easy matter, as the substances which are generally used to make the japan hard, have a tendency, by a number of coats, to look or become dull in brightness. One white ground is made by the following composition : white flake or lead washed over and ground up with a sixth of its weight of starch, then dried and mixed with the finest gum, ground up in parts of one ounce gum, to half an ounce of rectified turpentine mixed and ground thoroughly together.   This is to be finely laid on the article to be japanned, dried, and then varnished with five or six coats of the following: two ounces of the whitest seed-lac to three ounces of gum-anima reduced to a fine powder and dissolved in a quart of alcohol. This lac must be carefully picked. For a softer varnish than this, a little turpentine should be added, and less of the gum. A very good varnish and not brittle, may be made by dissolving gum-anima in nut oil, boiling it gently as the gum is added, and giving the oil as much gum as it will take up. The ground of white varnish may of itself be made of this varnish, by giving two or three coats of it, but when used it should be diluted with pure turpentine. Although this varnish is not brittle it is liable to be indented with strokes, and it will not bear to be polished, but if well laid on it will not need polishing afterwards ; it also takes some time to dry. Heat applied to all oils, however, darkens their color, and oil varnishes for white grow very yellow if not exposed to a full clear light.

GUM COPAL.

Copal varnish is one of the very finest varnishes for japanning purposes. It can be dissolved by linseed oil, rendered dry by adding some quicklime at a heat somewhat less than will boil or decompose the oil by it.

This solution, with the addition of a little turpentine, forms a very transparent varnish, which, when properly applied and slowly dried is very hard and durable. This varnish is applied to snuff boxes, tea boards and other utensils. It also preserves paintings and renders their surfaces capable of reflecting light more uniformly.

If powdered copal be mixed in a mortar with camphor, it softens and becomes a coherent mass, and if camphor be added to alcohol it becomes an excellent solvent of copal by adding the copal well ground, and employing a tolerable degree of heat, having the vessel well corked which must have a long neck for the allowance of expansion, and the vessel must only be about one-fourth filled with the mixture. Copal can also be incorporated with turpentine, with one part of powdered copal to twelve parts of pure turpentine, subjected to the heat of a sand-bath for several days in a long necked mattress, shaking it frequently.

Copal is a good varnish for metals, such as *tin ;* the varnish must be dried in an oven, each coat, and it can be colored with some substances, but alcohol varnish will mix with any coloring matter. For white japans or varnishes, we have already shown that fine chalk or white lead was used as a basis, and the varnishes coated over it.

To japan or varnish white leather, so that it may be elastic, is altogether a different work from varnishing or japanning wood or metal, or papier mache.

For white leather oil is the principal ingredient, as it is well known that chalk is extensively used to give white leather its pure color, or speaking more philosophically, its fair colorless whiteness. White leather having already the basis of white varnish, it should get a light coat of the pure varnish, before mentioned, and dried well *in the oven,* or a coat of the oil copal will answer very well. This being well dried, boiled nut oil nicely coated and successively dried, will make a most beautiful white varnish for leather, not liable to crack. This quality takes a long time to dry, and of course is more expensive. Coarse varnish may be made of boiled linseed oil, into which is added gradually the acetate of lead as a drier. This addition must be done very cautiously as the oil will be apt to foam over.

A better and more safe drying mixture than the mere acetate of lead, is, to dissolve the acetate of lead in a small quantity of water, neutralize the acid with the addition of pipe clay, evaporate the sediment to perfect dryness, and feed the oil when gently boiling gradually with it.

These *varnishes* or *japans,* as far as described, have only reference to white grounds.

There is some nice work to be observed, and there is much in applying the varnishes at the right time, knowing by the eye the proper moment when the mixture is perfect, or when to add any ingredient. These things require practice.

BLACK GROUNDS.

Black grounds for japans may be made by mixing ivory black with shellac varnish ; or for coarse work, lamp black and the top coating of common seedlac varnish. A common black japan may be made by painting a piece of work with drying oil, (oil mixed with lead,) and putting the work into a stove, not too hot, but of such a degree, gradually raising the heat and keeping it up for a long time, so as not to burn the oil and make it blister. This process makes very fair japan and requires no polishing.

BLACK JAPAN.

Naples asphaltum fifty pounds, dark gum-anime eight pounds, fuse, add linseed oil twelve gallons, boil, add dark gum amber ten pounds, previously fused and boiled with linseed oil two gallons, add the driers, and proceed as last. *Used* for wood or metals.

BRUNSWICK BLACK.

1. Foreign asphaltum forty-five pounds, drying oil six gallons, litharge six pounds, boil as last, and thin with twenty-five gallons of oil of turpentine. *Used* for ironwork, &c. 2. Black pitch and gas tar asphaltum, of each twenty-five pounds, boil gently for five hours, then add linseed oil eight gallons, litharge and red lead, of each ten pounds, boil as before, and thin with oil of turpentine twenty gallons. Inferior to the last, but cheaper.

BLUE JAPAN GROUNDS.

Blue japan grounds may be formed of bright Prussian blue. The color may be mixed with shellac varnish, and brought to a polishing state by five or six coats of varnish of seed-lac. The varnish, however, is apt to give a greenish tinge to the blue, as the varnish has a yellowish tinge, and blue and yellow form a green. Whenever a light blue is desired, the purest varnish must always be used.

SCARLET JAPAN.

Ground vermilion may be used for this, but being so glaring it is not beautiful unless covered over with rose-pink, or lake, which have a good effect when thus used. For a very bright crimson ground, safflower or Indian lake should be used, always dissolved in the alcohol of which the varnish is made. In place of this lake, carmine may be used, as it is more common. The top coat of varnish must always be of the white seed-lac, which has been before described, and as many coats given as will be thought proper ; it is easy to judge of this.

YELLOW GROUNDS.

If turmeric be dissolved in the spirit of wine and strained through a cloth, and then mixed with pure seed-lac varnish, it makes a good yellow japan. Saffron will answer for the same purpose in the same way, but the brightest yellow ground is made by a primary coat of pure crome yellow, and coated successively with the varnish.

Dutch pink is used for a kind of cheap yellow japan ground. If a little dragon's blood be added to the varnish for yellow japan, a most beautiful and rich salmon-colored varnish is the result, and by these two mixtures all the shades of flesh-colored japans are produced.

GREEN JAPAN GROUNDS.

A good green may be made by mixing Prussian blue along with the cromate of lead, or with turmeric, or orpiment, (sulphuret of arsenic) or ochre, only the two should be ground together and dissolved in alcohol and applied as a ground, then coated with four or five coats of shellac varnish, in the manner already described. A very bright green is made by laying on a ground of Dutch metal, or leaf of gold, and then coating it over with distilled verdigris dissolved in alcohol, then the varnishes on the top. This is a splendid green, brilliant and glowing.

ORANGE COLORED GROUNDS.

Orange grounds may be made of yellow mixed with vermilion or carmine, just as a bright or rather inferior color is wanted. The yellow should always be in quantity to make a good full color, and the red added in proportion to the depth of shade. If there is not a good full body of yellow, the color will look watery, or bare, as it is technically termed.

PURPLE JAPAN GROUNDS.

This is made by a mixture of lake and Prussian blue, or carmine, or for an inferior color vermilion, and treated as the foregoing. When the ground is laid on and perfectly dried, a fine coat of pure boiled nut oil then laid on and perfectly dried, is a good method to have a japan, not liable to crack. But a better plan is to use this oil in the varnish given, the first coat, after the ground is laid on, and which should contain considerable of pure turpentine. In every case where oil is used for any purpose for varnish, it is all the better if turpentine is mixed with it. Turpentine enables oils to mix with either alcohol or water. Alkalies have this property also.

BLACK JAPAN.

1. Asphaltum three ounces, boiled oil four quarts, burnt umber eight ounces. Mix by heat, and when cooling thin with turpentine. 2. Amber twelve ounces, asphaltum two ounces ; fuse by heat, add boiled oil half a pint, resin two ounces ; when cooling add sixteen ounces oil of turpentine. Both are used to varnish metals.

JAPAN BLACK FOR LEATHER.

1. Burnt umber four ounces, true asphaltum two ounces, boiled oil two quarts. Dissolve the asphaltum by heat in a little of the oil, add the burnt umber ground in oil, and the remainder of the oil, mix, cool, and thin with turpentine. Flexible. 2. Shellac one part, wood naphtha four parts, dissolve, and color with lampblack. Inflexible.

TRANSPARENT JAPAN.

Oil of turpentine four ounces, oil of lavender three ounces, camphor one-half drachm, copal one ounce ; dissolve. Used to japan *tin*, but quick copal varnish is mostly used instead.

JAPANNERS' COPAL VARNISH.

Pale African copal seven pounds, fuse, add clarified linseed oil one half gallon, boil for five minutes, remove it into the open air, add boiling oil of turpentine three gallons, mix well, strain it into the cistern, and cover it up immediately. *Used* to varnish furniture, and by japanners, coachmakers, &c. *Dries* in 15 minutes, and may be polished as soon as hard.

5*

TORTOISE SHELL JAPAN.

This varnish is prepared by taking of good linseed oil one gallon, and of umber half a pound, and boiling them together until the oil becomes very brown and thick, when they are strained through a cloth and boiled again until the composition is about the consistence of pitch, when it is fit for use. Having prepared this varnish, clean well the copper or iron plate or vessel that is to be varnished, (japanned,) and then lay vermillion, mixed with shellac varnish, or with drying oil, diluted with turpentine, very thinly on the places intended to imitate the clean parts of the tortoise shell. When the vermillion is dry brush over the whole with the above umber varnish diluted to a due consistence with turpentine, and when it is set and firm, it must be put into a stove and undergo a strong heat for a long time, even two weeks will not hurt it. This is the ground for those beautiful snuff boxes and tea boards which are so much admired, and those grounds can be decorated with all kinds of paintings that fancy may suggest, and the work is all the better to be finished in an annealing oven.

PAINTING JAPAN WORK.

The colors to be painted are tempered, generally, in oil, which should have at least one-fourth of its weight of gum sandarach, or mastic dissolved in it, and it should be well diluted with turpentine, that the colors may be laid on thin and evenly. In some instances it does well to put on water colors or grounds of gold, which a skilful hand can do and manage so as to make the work appear as if it was embossed. These water colors are best prepared by means of isinglass size, mixed with honey, or sugar candy. These colors when laid on must receive a number of upper coats of the varnish we have described before.

JAPANNING OLD TEA-TRAYS.

First clean them thoroughly with soap and water and a little rotten stone; then dry them by wiping and exposure at the fire. Now, get some good copal varnish, mix with it some bronze powder, and apply with a brush to the denuded parts. After which set the tea-tray in an oven at a heat of 212° or 300° until the varnish is dry. Two coats will make it equal to new.

JAPAN FINISHING.

The finishing part of japanning lies in laying on and polishing the outer coats of varnish, which is necessary in all painted or simply ground colored japan work. When brightness and clearness are wanted, the white kind of varnish is necessary, for seed-lac varnish, which is the hardest and most tenacious, imparts a yellow tinge. A mixed varnish, we believe, is the best for this purpose, that is, for combining hardness and purity. Take then three ounces of seed-lac, picked very carefully from all sticks and dirt and washing it well with cold water, stirring it up, pouring it off, and continuing the process until the water runs off perfectly pure. Dry it and then reduce it to powder, and put it with a pint of pure alcohol into a bottle, of which it must occupy only two-thirds of its space. This mixture must be shaken well together and the bottle kept at a gentle heat (being corked) until the lac be dissolved. When this is the case, the clear must be poured off, and the remainder strained through a cloth, and all the clear, strained and poured, must be kept in a well stopped bottle. The manner of using this seed-lac varnish is the same as that before described, and a fine polishing varnish is made by mixing this with the pure white varnish. The pieces of work to be varnished for finishing should be placed near a stove, or in a warm, dry room, and one coat should be perfectly dry before the other is applied. The varnish is applied by proper brushes, beginning at the middle, passing the stroke to one end and with the other stroke from the middle to the other end. Great skill is displayed in laying on these coats of varnish. If possible the skill of hand should never cross, or twice pass over in giving one coat. When one coat is dry another must be laid over it, and so on successively for a number of coats, so that the coating should be sufficiently thick to stand fully all the polishing, so as not to bare the surface of the colored work. When a sufficient number of coats are thus laid on, the work is fit to be polished, which, in common cases, is commenced with a rag dipped in finely powdered rotten stone, and towards the end of the rubbing a little oil should be used along with the powder, and when the work appears fine and glossy a little oil should be used alone to clean off the powder and give the work a still brighter hue. In very fine work, French whiting should be used, which should be washed in water to remove any sand that might be in it. Pumice stone ground to a very fine powder is used for the first part of polishing, and the finishing done with whiting. It is always best to dry the varnish of all japan work by heat. For wood work, heat must be sparingly used, but for metals the varnish should be dried in an oven, also for papier mache and leather. The metal will stand the greatest heat, and care must be taken not to darken by too high a temperature. When gold size is used in gilding for japan work, where it is desired not to have the gold shine, or appear burnished, the gold size should be used with a little of the spirits of turpentine and a little oil, but when a considerable degree of lustre is wanted without burnishing and the preparation necessary for it, a little of the size along with oil alone should be used.

VARNISHES, — MISCELLANEOUS.

Different substances are employed for making varnish, the object being to produce a liquid easily applied to the surface of cloth, paper or metal, which, when dry, will protect it with a fine skin.

Gums and resins are the substances employed for making varnishes; they are dissolved either in turpentine, alcohol, or oil, in a close stone ware, glass or metal vessel, exposed to a low heat, as the case may require, or cold. The alcohol or turpentine dissolves the gum or resin, and holds them in solution, and after the application of the varnish, this mixture being mechanical, the moisture of the liquid evaporates, and the gum adheres to the article to which it is applied.

———

The choice of linseed oil is of peculiar consequence to the varnish-maker. Oil from fine full-grown ripe seed, when viewed in a vial, will appear limpid, pale, and brilliant ; it is mellow and sweet to the taste, has very little smell, is specifically lighter than impure oil, and, when clarified, dries quickly and firmly, and does not materially change the color of the varnish when made, but appears limpid and brilliant.

———

The following are the chief Resins employed in the manufacture of Varnishes.

AMBER.

This resin is most distinguished for durability. It is usually of some shade of yellow, transparent, hard, and moderately tough. Heated in air, it fuses at about 549° ; it burns with a clear flame, emitting a pleasant odor.

ANIME.

This is imported from the East Indies. The large, transparent, pale-yellow pieces, with vitreous fracture, are best suited for varnish. Inferior qualities are employed for manufacturing gold-size or japan-black. Although superior to amber in its capacity for drying, and equal in hardness, varnish made from anime deepens in color on exposure to air, and is very liable to crack. It is, however, much used for mixing with copal varnish.

BENZOIN.

This is a gum-resin but little used in varnishes, on account of its costliness.

COLOPHONY.

This resin is synonymous with arcanson and rosin. When the resinous juice of *Pinus sylvestris* and other varieties is distilled, colophony remains in the retort. Its dark color is due to the action of the fire. Dissolved in linseed oil, or in turpentine by the aid of heat, colophony forms a brilliant, hard, but brittle varnish.

COPAL.

This is a gum-resin of immense importance to the varnish-maker. It consists of several minor resins of different degrees of solubility.

Iu durability, it is only second to amber. When made into varnish, the better sorts become lighter in color by exposure to air.

Copal is generally imported in large lumps about the size of potatoes. The clearest and palest are selected for what is called *body-gum;* the second best forms *carriage-gum;* whilst the residue, freed from the many impurities with which it is associated, constitutes *worst quality,* fitted only for japan-black or gold-size.

In alcohol, copal is but little soluble ; but it is said to become more so by reducing it to a fine powder, and exposing it to atmospheric influences for twelve months. Boiling alcohol or spirit of turpentine, when poured upon *fused* copal, accomplishes its complete solution, provided the solvent be not added in too large proportions at a time. The addition of camphor also promotes the solubility of copal ; so likewise does oil of rosemary.

DAMMARA.

This is a tasteless, inodorous, whitish resin, easily soluble in oils. It is not so hard as mastic, with which it forms a good admixture.

ELEMI.

This is a resin of a yellow color, semi-transparent, and of faint fragrance. Of the two resins which it contains, one is crystallizable and soluble in cold alcohol.

LAC.

This constitutes the basis of spirit-varnish. The resin is soluble in strong alcohol aided by heat. Its solution in ammonia may be used as a varnish, when the articles coated with it are not exposed more than an hour or two at a time to water.

MASTIC.

This is a soft resin of considerable lustre. The two sorts in commerce are, *in tears* and the *common mastic ;* the former is the purer of the two. It consists of two resins, one of which is soluble in dilute alcohol. With oil of turpentine, it forms a very pale varnish, of great lustre, which flows readily, and works easily. Moreover, it can be readily removed by friction with the hand ; hence its use for delicate work of every description.

SANDARACH.

This is a pale, odorous resin, less hard than lac, with which it is often associated as a spirit-varnish. It consists of three resins differing as to solubility in alcohol, ether, and turpentine. It forms a good pale varnish for light-colored woods ; when required to be polished, Venice turpentine is added to give it body.

Of the solvents of these various resins, little need be said. In the manufacture of varnishes, great care, as well as cleanliness, are required. The resins should be washed in hot water, to free them from particles of dust and dirt ; they should be dried and assorted according to their color, reserving the lightest shades for the best kinds of varnish.

The *linseed-oil* should be as pale colored, and as well clarified as possible. New oil always contains mucilage, and more or less of foreign matters ; as these prevent the regular absorption of oxygen, the oil requires preliminary treatment. The common plan is to boil it with litharge ; but such *oil varnish* is inferior to that prepared with sulphate of lead.

The best method is to rub up linseed-oil with dry sulphate of lead, in sufficient quantity to form a milky mixture. After a week's exposure to the light, and frequent shaking, the mucus deposits with the sulphate of lead, and leaves the oil perfectly clear. The precipitated slime forms a compact membrane over the lead, hardening to such an extent that the clarified oil may be readily poured off.

TURPENTINE.

This is of very extensive use. The older it is, the more ozonized, the better it is. Turpentine varnishes dry much more readily than oil varnishes, are of a lighter color, more flexible and cheap. They are, however, neither so tough nor so durable.

ALCOHOL.

This is employed as the solvent of sandarach and of lac. The stronger, *cæteris paribus*, the better.

NAPHTHA AND METHYLATED SPIRIT OF WINE.

These are used for the cheaper varnishes. Their smell is disagreeable. The former is, however, a better solvent of resins than alcohol.

SPIRIT VARNISHES.

These varnishes may be readily colored—*red*, by dragon's blood ; *yellow*, by gamboge. If a colored varnish is required, clearly no account need be taken of the color of the resins. Lac varnish may be bleached by Mr. Lemming's process :— Dissolve five ounces of shellac in a quart of spirit of wine ; boil for a few minutes with ten ounces of well-burnt and recently-heated animal charcoal, when a small quantity of the solution should be drawn off and filtered : if not colorless. a little more charcoal should be added. When all tinge is removed, press the liquor through silk, as linen absorbs more varnish ; and afterwards filter it through fine blotting-paper. Dr. Hare proceeds as follows :—Dissolve in an iron kettle about one part of pearlash in about eight parts of water, add one part of shell or seed lac, and heat the whole to ebullition. When the lac is dissolved, cool the solution, and impregnate it with chlorine gas till the lac is all precipitated. The precipitate is white, but the color deepens by washing and consolidation. Dissolved in alcohol, lac bleached by this process yields a varnish which is as free from color as any copal varnish.

One word in conclusion with reference to all spirit varnishes. A damp atmosphere is sufficient to occasion a milky deposit of resin, owing to the diluted spirit depositing a portion : in such case the varnish is said to be *chilled*.

ESSENCE VARNISHES.

They do not differ essentially in their manufacture from spirit varnishes. The polish produced by them is more durable, although they take a longer time to dry.

OIL VARNISHES.

The most durable and lustrous of varnishes are composed of a mixture of resin, oil, and spirit of turpentine. The oils most frequently employed are linseed and walnut ; the resins chiefly copal and amber.

The drying power of the oil having been increased by litharge, red-lead, or by sulphate of lead, and a judicious selection of copal having been made, it is necessary, according to Booth, to bear in mind the following precautions before proceeding to the manufacture of varnish :—1. That oil varnish is not a solution, but an intimate mixture of resin in boiled oil and spirit of turpentine. 2. That the resin must be completely fused previous to the addition of the boiled or prepared oil. 3. That the oil must be heated from 250° to 300°. 4. That the spirit of turpentine must be added gradually, and in a thin stream, while the mixture of oil and resin is still hot. 5. That the varnish be made in dry weather, otherwise moisture is absorbed, and its transparency and drying quality impaired.

The heating vessel must be of copper, with a riveted and not a soldered bottom. To promote the admixture of the copal with the *hot* oil, the copal—carefully selected, and of nearly uniform fusibility—is *separately* heated with continuous stirring over a charcoal fire. Good management is required to prevent the copal from burning or becoming even high colored. When completely fused, the heated oil should be gradually poured in with constant stirring. The *exact* amount of oil required must be determined by experiment. If a drop upon a plate, on cooling, assumes such a consistency as to be penetrated by the nail without cracking, the mixture is complete ; but if it cracks, more oil must be added.

The spirit of turpentine *previously heated* is added in a thin stream to the former mixture, care being taken to keep up the heat of all the parts.

LACKER.

This is used for wood or brass work, and is also a varnish. For brass, the proportions are half a pound of pale shell-lac to one gallon of spirit of wine. It is better prepared without the aid of heat, by simple and repeated agitation. It should then be left to clear itself, and separated from the thicker portions and from all impurities by decantation. As it darkens on exposure to light, the latter should be excluded. It need scarcely be said that the color will be also modified by that of the lac employed.

1. COPAL VARNISHES.

1. Oil of turpentine one pint, set the bottle in a water bath, and add in small portions at a time, three ounces of powdered copal that has been previously melted by a gentle heat, and dropped into water ; in a few days decant the clear. *Dries* slowly, but is very pale and durable. *Used* for pictures, &c. 2. Pale hard copal two pounds ; fuse, add hot drying oil one pint, boil as before directed, and thin with oil of turpentine three pints, or as much as sufficient. Very pale. *Dries* hard in 12 to 24 hours. 3. Clearest and palest African copal eight pounds ; fuse, add hot and pale drying oil two gallons, boil till it strings strongly, cool a little, and thin with hot rectified oil of turpentine three gallons, and immediately strain into the store can. Very fine. Both the above are used for pictures. 4. Coarsely-powdered copal and glass, of each four ounces, alcohol of 90 per cent one pint, camphor one-half ounce ; heat it in a water-bath so that the bubbles may be counted as they rise, observing frequently to stir the mixture ; when cold decant the clear. *Used* for pictures. 5. Copal melted and dropped into water three ounces, gum sandarach six ounces, mastic and Chio turpentine of each two and one-half ounces, powdered glass four ounces, alcohol of 85 per cent, one quart ; dissolve by a gentle heat. *Used* for metal, chairs, &c.

All copal varnishes are hard and durable, though less so than those made of amber, but they have the advantage over the latter of being paler. They are applied on coaches, pictures, polished metal, wood, and other objects requiring good durable varnish.

2. COPAL VARNISH.

Hard copal, 300 parts ; drying linseed or nut oil, from 125 to 250 parts ; oil of turpentine, 500 ; these three substances are to be put into three separate vessels ; the copal is to be fused by a somewhat sudden application of heat; the drying oil is to be heated to a temperature a little under ebullition, and is to be added by small portions at a time to the melted copal. When this combination is made, and the heat a little abated, the essence of turpentine, likewise previously heated, is to be introduced by degrees ; some of the volatile oil will be dissipated at first, but more being added, the union will take place. Great care must be taken to prevent the turpentine vapor from catching fire, which might occasion serious accidents to the operator. When the varnish is made and has cooled down to about 130 degrees of Fah., it may be strained through a filter, to separate the impurities and undissolved copal. Almost all varnish makers think it indispensable to combine the drying oil with the copal before adding the oil of turpentine, but in this they are mistaken. Boiling oil of turpentine combines very readily with fused copal; and, in some cases, it would probably be preferable to commence the operation with it, adding it in successive small quantities. Indeed, the whitest copal varnish can be made only in this way ; for if the drying oil has been heated to nearly its boiling point, it becomes colored, and darkens the varnish.

This varnish improves in clearness by keeping. Its consistence may be varied by varying the proportions of the ingredients within moderate limits. Good varnish, applied in summer, should become so dry in twenty-four hours that the dust will not stick to it nor receive an impression from the fingers. To render it sufficiently dry and hard for polishing, it must be subjected for several days to the heat of a stove.

3. COPAL VARNISHES.

1. Melt in an iron pan at a slow heat, copal gum, powdered, eight parts, and add balsam copaiva, previously warmed, two parts. Then remove from the fire, and add spirits of turpentine, also warmed beforehand, ten parts, to give the necessary consistence. 2. Prepared gum copal ten parts, gum mastic two parts, finely powdered, are mixed with white turpentine and boiled linseed oil, of each one part, at a slow heat, and with spirits of turpentine twenty parts. 3. Prepared gum-copal ten parts, white turpentine two parts, dissolve in spirits of turpentine.

Gum-copal is *prepared* or made more soluble in spirits of turpentine, by melting the powdered crude gum, afterwards again powdering, and allowing to stand for some time loosely covered.

CABINET VARNISH.

Copal, fused, fourteen pounds; linseed oil, hot, one gallon; turpentine, hot, three gallons. Properly boiled, such a varnish will dry in ten minutes.

TABLE VARNISH

Damma resin, one pound; spirits of turpentine, two pounds; camphor, two hundred grains. Digest the mixture for twenty-four hours. The decanted portion is fit for immediate use.

COMMON TABLE VARNISH.

Oil of turpentine, one pound; bees' wax, two ounces; colophony, one drachm.

COPAL VARNISH FOR INSIDE WORK.

1. Pounded and oxidixed copal, twenty-four parts; spirit of turpentine, forty parts; camphor, one part.—2. *Flexible Copal Varnish.* Copal in powder, sixteen parts; camphor, two parts; oil of lavender, ninety parts,

Dissolve the camphor in the oil, heat the latter, and stir in the copal in successive portions until complete solution takes place. Thin with sufficient turpentine to make it of proper consistence.

BEST BODY COPAL VARNISH FOR COACH MAKERS, &c.

This is intended for the body parts of coaches and other similar vehicles, intended for polishing. Fuse eight lbs. of fine African gum copal, and two gallons of clarified oil, boil it very slowly for four or five hours, until quite stringy, mix with three gallons and a

half of turpentine ; strain off and pour it into a cistern. If this is
too slow in drying, coach-makers, painters and varnish-makers have
introduced to two pots of the preceding varnish, one made as follows :
eight lbs. of fine pale gum-anime, two gallons of clarified oil and
three and a half gallons of turpentine. To be boiled four hours.

COPAL POLISH.

Digest or shake finely powdered gum copal four parts, and gum
camphor one part, with ether to form a semi-fluid mass, and then
digest with a sufficient quantity of alcohol.

WHITE SPIRIT VARNISH.

Sandarach, 250 parts ; mastic, in tears, 64 ; elemi resin, 82 ;
turpentine, 64 ; alcohol of 85 per cent, 1000 parts, by measure.
The turpentine is to be added after the resins are dissolved. This is
a brilliant varnish, but not so hard as to bear polishing.   .

WHITE HARD SPIRIT VARNISHES.

1. Gum sandarach five pounds, camphor one ounce, rectified spirit
(65 over proof) two gallons, washed and dried coarsely-pounded glass
two pounds ; proceed as in making mastic varnish ; when strained
add one quart of very pale turpentine varnish. Very fine. 2. Picked
mastic and coarsely-ground glass, of each, four ounces, sandarach
and pale clear Venice turpentine, of each three ounces, alcohol two
pounds ; as last. 3. Gum sandarach one pound, clear Strasburgh
turpentine six ounces, rectified spirit (65 over proof) three pints ;
dissolve. 4. Mastic in tears two ounces, sandarach eight ounces, gum
elemi one ounce, Strasburgh or Scio turpentine (genuine) four ounces,
rectified spirit (65 over proof) one quart. *Used* on metals, &c.
Polishes well.

WHITE VARNISH.

1. Tender copal seven and one-half ounces, camphor one ounce,
alcohol of 95 per cent, one quart ; dissolve, then add mastic two
ounces, Venice turpentine one ounce ; dissolve and strain. Very
white, drying, and capable of being polished when hard. *Used* for
toys. 2. Sandarach eight ounces, mastic two ounces, Canada balsam
four ounces, alcohol one quart. *Used* on paper, wood, or linen.

SOFT BRILLIANT VARNISH.

Sandarach six ounces, elemi (genuine) four ounces, anime one
ounce, camphor one-half ounce, rectified spirit one quart ; as before.

The above spirit varnishes are chiefly applied to objects of the toil-
ette, as work-boxes, card-cases, &c., but are also suitable to other
articles, whether of paper, wood, linen, or metal, that require a brilliant and quick-drying varnish. They mostly dry almost as soon as
applied, and are usually hard enough to polish in 24 hours. Spirit
varnishes are less durable and more liable to crack than oil varnishes.

BROWN HARD SPIRIT VARNISHES.

1. Sandarach four ounces, pale seed-lac two ounces, elemi (true) one ounce, alcohol one quart ; digest with agitation till dissolved, then add Venice turpentine two ounces. 2. Gum sandarach three pounds, shellac two pounds, rectified spirit, (65 over proof,) two gallons ; dissolve, add turpentine varnish one quart ; agitate well and strain. Very fine. 3. Seed-lac and yellow resin, of each one and one-half pounds, rectified spirit two gallons.

TO PREPARE A VARNISH FOR COATING METALS.

Digest one part of bruised copal in two parts of absolute alcohol; but as this varnish dries too quickly it is preferable to take one part of copal, one part of oil of rosemary, and two or three parts of absolute alcohol. This gives a clear varnish as limped as water. It should be applied hot, and when dry it will be found hard and durable.

TO VARNISH ARTICLES OF IRON AND STEEL.

Dissolve 10 parts of clear grains of mastic, 5 parts of camphor, 15 parts of sandarach, and 5 of elemi, in a sufficient quantity of alcohol, and apply this varnish without heat. The articles will not only be preserved from rust, but the varnish will retain its transparency and the metallic brilliancy of the articles will not be obscured.

VARNISH FOR IRON WORK.

Dissolve, in about two lbs. of tar oil, half a pound of asphaltum, and a like quantity of pounded resin, mix hot in an iron kettle, care being taken to prevent any contact with the flame. When cold the varnish is ready for use. This varnish is for out-door wood and iron work, not for japanning leather or cloth.

BLACK VARNISH FOR IRON WORK.

Asphaltum forty-eight pounds, fuse, add boiled oil ten gallons, red lead and litharge, of each seven pounds, dried and powdered white copperas three pounds, boil for two hours, then add dark gum amber (fused) eight pounds, hot linseed oil two gallons, boil for two hours longer, or till a little of the mass, when cooled, may be rolled into pills, then withdraw the heat, and afterwards thin down with oil of turpentine thirty gallons. *Used* for the ironwork of carriages, and other nice purposes.

BRONZE VARNISH FOR STATUARY.

Cut best hard soap fifty parts, into fine shavings, dissolve in boiling water two parts, to which add the solution of blue vitriol fifteen parts, in pure water sixty parts. Wash the copper-soap with water, dry it at a very slow heat, and dissolve it in spirits of turpentine.

AMBER VARNISHES.

1. Amber one pound, pale boiled oil ten ounces, turpentine one pint. Render the amber, placed in an iron pot, semiliquid by heat; then add the oil, mix, remove it from the fire, and when cooled a a little, stir in the turpentine. 2. To the amber, melted as above, add two ounces of shellac, and proceed as before.

This varnish is rather dark, but remarkably tough. The first form is the best. It is used for the same purposes as copal varnish, and forms an excellent article for covering wood, or any other substance not of a white or very pale color. It dries well, and is very hard and durable.

AMBER VARNISH, BLACK.

Amber one pound, boiled oil one-half pint, powdered asphaltum six ounces, oil of turpentine one pint. Melt the amber, as before described, then add the asphaltum, previously mixed with the cold oil, and afterwards heated very hot, mix well, remove the vessel from the fire, and when cooled a little add the turpentine, also made warm.

Each of the above varnishes should be reduced to a proper consistence with more turpentine if required. The last form produces the *beautiful black varnish* used by the coachmakers. Some manufacturers omit the whole or part of the asphaltum, and use the same quantity of clear black rosin instead, in which case the color is brought up by lampblack reduced to an impalpable powder, or previously ground very fine with a little boiled oil. The varnish made in this way, lacks, however, that richness, brilliancy, and depth of blackness imparted by asphaltum.

AMBER VARNISHES.

1. (*Pale.*) Amber pale and transparent six pounds, fuse, add hot clarified linseed oil two gallons, boil till it strings strongly, cool a little, and add oil of turpentine four gallons. Pale as copal varnish; soon becomes very hard, and is the most durable of oil varnishes; but requires time before it is fit for polishing. When wanted to dry and harden quicker, "drying" oil may be substituted for linseed, or "driers" may be added during the boiling. 2. Amber one pound; melt, add Scio turpentine one-half pound, transparent white resin two ounces, hot linseed oil one pint, and afterwards oil of turpentine as much as sufficient; as above. Very tough. 3. (*Hard.*) Melted amber four ounces, hot boiled oil one quart; as before. 4. (*Pale.*) Very pale and transparent amber four ounces, clarified linseed oil and oil of turpentine, of each one pint; as before.

Amber varnish is suited for all purposes, where a very hard and durable oil varnish is required. The paler kind is superior to copal varnish, and is often mixed with the latter to increase its hardness and durability.

BLACK VARNISH.

Heat to boiling linseed oil varnish ten parts, with burnt umber two parts, and powdered asphaltum one part, and when cooled dilute with spirits of turpentine to the required consistence.

VARNISH FOR CERTAIN PARTS OF CARRIAGES.

Sandarach, 190 parts ; pale shellac, 95 ; resin, 125 ; turpentine, 190 ; alcohol, at 85 per cent, 1000 parts, by measure.

COACH VARNISH.

Mix shellac sixteen parts, white turpentine three parts, lampblack sufficient quantity, and digest with alcohol ninety parts, oil of lavender four parts.

MAHOGANY VARNISH.

Sorted gum-anime eight pounds, clarified oil three gallons, litharge and powdered dried sugar of lead, of each one-fourth pound ; boil till it strings well, then cool a little, thin with oil of turpentine five and one-half gallons, and strain.

VARNISH FOR CABINET MAKERS.

Pale shellac, 750 parts ; mastic, 64 ; alcohol, of 90 per cent, 1000 parts by measure. The solution is made in the cold, with the aid of frequent stirring. It is always muddy, and is employed without being filtered. With the same resins and proof spirit a varnish is made for the bookbinders to do over their morocco leather.

CEMENT VARNISH FOR WATER-TIGHT LUTING.

White turpentine fourteen parts, shellac eighteen parts, resin six parts, digest with alcohol eighty parts.

THE VARNISH OF WATIN FOR GILDED ARTICLES.

Gum-lac, in grain, 125 parts ; gamboge, 125 ; dragon's blood, 125 ; annotto, 125 ; saffron, 32. Each resin must be dissolved in 1000 parts by measure, of alcohol of 90 per cent ; two separate tinctures must be made with the dragon's blood and annotto, in 1000 parts of such alcohol ; and a proper proportion of each should be added to the varnish, according to the shade of golden color wanted.

CHEAP OAK VARNISH.

Clear pale resin three and one-half pounds, oil of turpentine one gallon ; dissolve. It may be colored darker by adding a little fine lampblack.

VARNISH FOR WOOD-WORK.

Powdered gum sandarach eight parts, gum mastic two parts, seed-lac eight parts, and digest in a warm place for some days with alcohol twenty-four parts, and finally, dilute with sufficient alcohol to the required consistence.

DARK VARNISH FOR LIGHT WOOD-WORK.

Pound up and digest shellac sixteen parts, gum sandarach thirty-two parts, gum mastic (juniper) eight parts, gum elemi eight
6*

parts, dragon's blood four parts, annotto one part, with white turpentine sixteen parts, and alcohol two hundred and fifty-six. Dilute with alcohol if required.

VARNISH FOR INSTRUMENTS.

Digest seed-lac one part, with alcohol seven parts, and filter.

VARNISH FOR THE WOOD TOYS OF SPA.

Tender copal, 75 parts ; mastic, 12.5 ; Venice turpentine, 6.5 ; alcohol, of 95 per cent, 100 parts by measure ; water ounces, for example, if the other parts be taken in ounces. The alcohol must be first made to act upon the copal, with the aid of a little oil of lavender or camphor, if thought fit ; and the solution being passed through a linen cloth, the mastic must be introduced. After it is dissolved, the Venice turpentine, previously melted in a water-bath, should be added ; the lower the temperature at which these operations are carried on, the more beautiful will the varnish be. This varnish ought to be very white, very drying, and capable of being smoothed with pumice-stone and polished.

VARNISHES FOR FURNITURE.

The simplest, and perhaps the best, is the solution of shellac only, but many add gums sandarach, mastic, copal, arabic, benjamin, &c., from the idea that they contribute to the effect. Gum arabic is certainly never required if the solvent be pure, because it is insoluble in either rectified spirit or rectified wood naphtha, the menstrua employed in dissolving the gums. As spirit is seldom used on account of its expense, most of the following are mentioned as solutions in naphtha, but spirit can be substituted when thought proper.
1. Shellac one and a half pounds, naphtha one gallon ; dissolve, and it is ready without filtering. 2. Shellac twelve ounces, copal three ounces, (or an equivalent of varnish); dissolve in one gallon of naphtha. 3. Shellac one and a half pounds, seed-lac and sandarach each four ounces, mastic two ounces, rectified spirit one gallon ; dissolve. 4. Shellac two pounds, benzoin four ounces, spirit one gallon. 5. Shellac ten ounces, seed-lac, sandarach, and copal varnish of each, six ounces, benzoin three ounces, naphtha one gallon.
To darken polish, benzoin and dragon's-blood are used, turmeric and other coloring matters are also added ; and to make it lighter it is necessary to use bleached lac, though some endeavor to give this effect by adding oxalic acid to the ingredients, it, like gum arabic, is insoluble in good spirit or naphtha. For all ordinary purposes the first form is best and least troublesome, while its appearance is equal to any other.

TO FRENCH POLISH.

The wood must be placed level, and sand-papered until it is *quite smooth*, otherwise it will *not polish*. Then provide a rubber of cloth, list, or sponge, wrap it in a soft rag, so as to leave a handle at the back for your hand, shake the bottle against the rubber, and in the middle of the varnish on the rag place with your finger a little raw linseed oil. Now commence rubbing, in small circular strokes, and continue until the pores are filled, charging the rubber with varnish and oil as required, until the whole wood has had one coat. When dry repeat the process once or twice until the surface appears even and fine, between each coat using fine sand-paper to smooth down all irregularities. Lastly, use a clean rubber with a little strong alcohol only, which will remove the oil and the cloudiness it causes ; when the work will be complete.

FURNITURE POLISHES.

New wood is often French-polished. Or the following may be tried : Melt three or four pieces of sandarach, each the size of a walnut, add one pint of boiled oil, and boil together for one hour. While cooling add one drachm of venice turpentine, and if too thick a little oil of turpentine also. Apply this all over the furniture, and after some hours rub it off ; rub the furniture daily, without applying fresh varnish, except about once in two months. Water does not injure this polish, and any stain or scratch may be again covered, which cannot be done with French-polish.

FURNITURE GLOSS.

To give a gloss to household furniture, various compositions are used, known as wax, polish, creams, pastes, oils, &c. The following are some of the forms used :

FURNITURE CREAM.

Bees-wax one pound, soap four ounces, pearlash two ounces, soft water one gallon ; boil together until mixed.

FURNITURE OILS.

1. Acetic acid two drachms, oil of lavender one-half drachm, rectified spirit one drachm, linseed oil four ounces. 2. Linseed oil one pint, alkanet root two ounces ; heat, strain, and add lac varnish one ounce. 3. Linseed oil one pint, rectified spirit two ounces, butter of antimony four ounces.

FURNITURE PASTES.

1. Bees-wax, spirit of turpentine, and linseed oil, equal parts ; melt and cool. 2. Bees-wax four ounces, turpentine ten ounces, alkanet root to color ; melt and strain. 3. Bees-wax one pound, linseed oil five ounces, alkanet root one-half ounce ; melt, add five ounces of turpentine, strain and cool. 4. Bees-wax four ounces, resin one ounce, oil of turpentine two ounces, venetian red to color.

ETCHING VARNISHES.

1. White wax, two ounces ; black and Burgundy pitch, of each one-half ounce ; melt together, add by degrees powdered asphaltum two ounces, and boil till a drop taken out on a plate will break when cold by being bent double two or three times between the fin-

gers ; it must then be poured into warm water and made into small balls for use. 2. (*Hard Varnish.*) Linseed oil and mastic, of each four ounces ; melt together. 3. (*Soft Varnish.*) Soft linseed oil, four ounces ; gum benzoin and white wax, of each one-half ounce ; boil to two-thirds.

VARNISH FOR ENGRAVINGS, MAPS, ETC.

Digest gum sandarach twenty parts, gum mastic eight parts, camphor one part, with alcohol forty-eight parts. The map or engraving must previously receive one or two coats of gelatine.

VARNISH TO FIX ENGRAVINGS OR LITHOGRAPHS ON WOOD.

For fixing engravings or lithographs upon wood, a varnish called mordant is used in France, which differs from others chiefly in containing more Venice turpentine, to make it sticky ; it consists of sandarach, 250 parts ; mastic in tears, 64 ; rosin, 125 ; Venice turpentine, 250 ; alcohol, 1000 parts by measure.

VARNISHES FOR OIL PAINTINGS AND LITHOGRAPHS.

1. Dextrine 2 parts, alcohol 1 part, water 6 parts. 2. Varnish for drawings and lithographs : dextrine 2 parts, alcohol ½ part, water 2 parts. These should be prepared previously with two or three coats of thin starch or rice boiled and strained through a cloth.

VARNISH FOR OIL PAINTINGS.

Digest at a slow heat gum sandarach two parts, gum mastic four parts, balsam copaiva two parts, white turpentine three parts, with spirits of turpentine four parts, alcohol (95 per cent) 50-56 parts.

BEAUTIFUL VARNISH FOR PAINTINGS AND PICTURES.

Honey, 1 pint; the whites of two dozen fresh hen's eggs; 1 ounce of good clean isinglass, 20 grains of hydrate of potassium, ½ ounce of chloride of sodium; mix together over a gentle heat of 80 or 90 degrees Fah.; be careful not to let the mixture remain long enough to coagulate the albumen of the eggs ; stir the mixture thoroughly, then bottle. It is to be applied as follows : one table spoonful of the varnish added to half a table spoonful of good oil of turpentine, then spread on the picture as soon as mixed.

MILK OF WAX.

Milk of wax is a valuable varnish, which may be prepared as follows :—Melt in a porcelain capsule a certain quantity of white wax, and add to it, while in fusion, an equal quantity of spirit of wine, of sp. grav. 0·830 ; stir the mixture, and pour it upon a large porphyry slab. The granular mass is to be converted into a paste by the muller, with the addition, from time to time, of a little alcohol ; and as soon as it appears to be smooth and homogeneous, water is to be introduced in small quantities successively, to the amount of four times the weight of the wax. This emulsion is to be then passed through canvas, in order to separate such particles as may be imperfectly incorporated. The milk of wax, thus prepared, may be spread with a smooth brush upon the surface of a painting, allowed to dry, and then fused by passing a hot iron (salamander) over its surface. When cold, it is to be rubbed with a linen cloth to bring out the lustre. It is to the unchangeable quality of an encaustic of this nature, that the ancient paintings upon the walls of Herculaneum and Pompeii owe their freshness at the present day.

CRYSTAL VARNISHES.

1. Genuine pale Canada balsam and rectified oil of turpentine, equal parts ; mix, place the bottle in warm water, agitate well, set it aside, in a moderately warm place, and in a week pour off the clear. *Used* for maps, prints, drawings, and other articles of paper, and also to prepare tracing paper, and to transfer engravings. 2. Mastic three ounces, alcohol one pint ; dissolve. *Used* to fix pencil drawings.

ITALIAN VARNISHES.

1. Boil Scio turpentine till brittle, powder, and dissolve in oil of turpentine. 2. Canada balsam and clear white resin, of each six ounces, oil of turpentine one quart ; dissolve. *Used* for prints, &c.

WATER VARNISH FOR OIL-PAINTINGS.

Boil bitter-apple, freed from the seeds and cut five parts, with rain-water fifty parts, down to one-half. Strain and dissolve in the liquor gum arabic eight parts, and rock-candy four parts, and lastly, add one part of alcohol. Let it stand for some days, and filter.

VARNISH FOR PAPER-HANGINGS.

Sandarach, four parts, mastic, seed-lac, white turpentine, of each two parts, gum elemi one part, alcohol twenty-eight parts. Digest with frequent shaking, and filter. Before applying this varnish, the paper must be twice painted over with a solution of white gelatine, and dried.

BOOK-BINDERS' VARNISH.

Shellac eight parts, gum benzoin three parts, gum mastic two parts, bruise, and digest in alcohol forty-eight parts, oil of lavender one-half part. Or, digest shellac four parts, gum mastic two parts, gum dammar and white turpentine of each one part, with alcohol (95 per cent) twenty-eight parts.

TO VARNISH CARDWORK.

Before varnishing cardwork, it must receive two or three coats of size, to prevent the absorption of the varnish, and any injury to the design. The size may be made by dissolving a little isinglass in hot water, or by boiling some parchment cuttings until dissolved. In either case the solution must be strained through a piece of clean muslin, and for very nice purposes, should be clarified with a little white of egg. A small clean brush, called by painters a sash tool, is the best for applying the size, as well as the varnish. A light delicate touch must be adopted, especially for the first coat, least the ink or colors be started, or smothered.

SIZE, OR VARNISH, FOR PRINTERS, ETC.

Best pale glue and white curd soap, of each 4 ounces ; hot water 3 pints ; dissolve, then add powdered alum 2 ounces. *Used* to size prints and pictures before coloring them.

VARNISH FOR BRICK WALLS.

A varnish made with one pound of sulphur boiled for half an hour in an iron vessel is a perfect protection from damp to brick walls. It should be applied with a brush, while warm.

MASTIC VARNISHES.

1. (*Fine.*) Very pale and picked gum mastic five pounds, glass pounded as small as barley, and well washed and dried two and one-half pounds, rectified turpentine two gallons ; put them into a clean four gallon stone or tin bottle, bung down securely, and keep rolling it backwards and forwards pretty smartly on a counter or any other solid place for at least four hours ; when, if the gum is all dissolved, the varnish may be decanted, strained through muslin into another bottle, and allowed to settle. It should be kept for six or nine months before use, as it thereby gets both tougher and clearer. 2. (*Second Quality.*) Mastic eight pounds, turpentine four gallons ; dissolve by a gentle heat, and add pale turpentine varnish one-half gallon. 3. Gum mastic six ounces, oil of turpentine one quart ; dissolve.

Mastic varnish is used for pictures, &c. ; when good, it is tough, hard, brilliant, and colorless. Should it get " *chilled*," one pound of well-washed silicious sand should be made moderately hot, and added to each gallon, which must then be well agitated for five minutes, and afterwards allowed to settle.

INDIA-RUBBER VARNISHES.

1. Cut up one pound of India rubber into small pieces and diffuse in half a pound of sulphuric ether, which is done by digesting in a glass flask on a sand bath. Then add one pound pale linseed oil varnish, previously heated, and after settling, one pound of oil of turpentine, also heated beforehand. Filter, while yet warm, into bottles. Dries slowly.

2. Two ounces India rubber finely divided and digested in the same way, with a quarter of a pound of camphene, and half an ounce of naphtha or benzole. When dissolved add one ounce of copal varnish, which renders it more durable. Principally for gilding.

3. In a wide mouthed glass bottle, digest two ounces of India rubber in fine shavings, with one pound of oil of turpentine, during two days, without shaking, then stir up with a wooden spatula. Add

another pound of oil of turpentine, and digest, with frequent agitation, until all is dissolved. Then mix a pound and a half of this solution with two pounds of very white copal-oil varnish, and a pound and a half of well boiled linseed oil, shake and digest in a sand bath, until they have united into a good varnish.—For morocco leather.

4. Four ounces India-rubber in fine shavings are dissolved in a covered jar by means of a sand bath, in two pounds of crude benzole, and then mixed with four pounds of hot linseed oil varnish, and a half pound of oil of turpentine. Dries very well.

5. *Flexible Varnish.*—Melt one pound of rosin, and add gradually half a pound of India rubber in very fine shavings, and stir until cold. Then heat again, slowly, add one pound of linseed oil varnish, heated, and filter.

6. *Another.*—Dissolve one pound of gum dammar, and a half pound of India rubber, in very small pieces, in one pound of oil of turpentine, by means of a water bath. Add one pound of hot oil varnish and filter.

7. India rubber in small pieces, washed and dried, are fused for three hours in a close vessel, on a gradually heated sand bath. On removing from the sand bath, open the vessel and stir for ten minutes, then close again, and repeat the fusion on the following day, until small globules appear on the surface. Strain through a wire sieve.

8. *Varnish for Waterproof Goods.*—Let a quarter of a pound of India rubber, in small pieces, soften in a half pound of oil of turpentine, then add two pounds of boiled oil, and let the whole boil for two hours over a slow coal fire. When dissolved, add again six pounds of boiled linseed oil and one pound of litharge, and boil until an even liquid is obtained. It is applied warm.

9. *Gutta Percha Varnish.*—Clean a quarter of a pound of Gutta Percha in warm water from adhering impurities, dry well, dissolve in one pound of rectified rosin oil, and add two pounds of linseed oil varnish, boiling hot. Very suitable to prevent metals from oxidation.

BLACK VARNISH FOR HARNESS.

Digest shellac twelve parts, white turpentine five parts, gum sandarach two parts, lampblack one part, with spirits of turpentine four parts, alcohol ninety-six parts.

BOILED OIL OR LINSEED-OIL VARNISH.

Boil linseed oil sixty parts, with litharge two parts, and white vitriol one part, each finely powdered, until all water is evaporated. Then set by. Or, rub up borate of manganese four parts, with some of the oil, then add linseed oil three thousand parts, and heat to boiling.

DAMMAR VARNISH.

Gum dammar ten parts, gum sandarach five parts, gum mastic one part, digest at a low heat, occasionally shaking, with spirits of turpentine twenty parts. Finally, add more spirits of turpentine to give the consistence of syrup.

COMMON VARNISH.

Digest shellac one part, with alcohol seven or eight parts.

WATERPROOF VARNISHES.

Take one pound of flowers of sulphur and one gallon of linseed oil, and boil them together until they are thoroughly combined. This forms a good varnish for waterproof textile fabrics. Another is made with four pounds oxyde of lead, twopounds of lampblack, five ounces of sulphur, and ten pounds of India rubber dissolved in turpentine. These substances, in such proportions, are boiled together until they are thoroughly combined. Coloring matters may be mixed with them. Twilled cotton may be rendered waterproof by the application of the oil sulphur varnish. It should be applied at two or three different times, and dried after each operation.

VARNISHES FOR BALLOONS, GAS BAGS, ETC.

1. India rubber in shavings one ounce ; mineral naphtha two lbs.; digest at a gentle heat in a close vessel till dissolved, and strain. 2. Digest one pound of Indian rubber, cut small, in six pounds oil of turpentine for 7 days, in a warm place. Put the mixture in a water bath, heat until thoroughly mixed, add one gallon of warm boiled drying oil, mix, and strain when cold. 3. Linseed oil one gallon ; dried white copperas and sugar of lead, each three ounces; litharge eight ounces ; boil with constant agitation till it strings well, then cool slowly and decant the clear. If too thick, thin it with quicker drying linseed oil.

GOLD VARNISH.

Digest shellac sixteen parts, gum sandarach, mastic, of each three parts, crocus one part, gum gamboge two parts, all bruised, with alcohol one hundred forty-four parts. Or, digest seed-lac, sandarach, mastic, of each eight parts, gamboge two parts, dragon's blood one part, white turpentine six parts, turmeric four parts, bruised, with alcohol one hundred twenty parts.

WAINSCOT VARNISH FOR HOUSE PAINTING AND JAPANNING.

Anime eight pounds ; clarified linseed oil three gallons ; litharge one-fourth pound ; acetate of lead one-half pound ; sulphate of copper one-fourth pound.

All these materials must be carefully but thoroughly boiled together until the mixture becomes quite stringy, and then five and a half gallons of heated turpentine stirred in. It can be easily deepened in color by the addition of a little gold-size.

LACKERS.

GOLD LACKER.

Put into a clean four gallon tin, one pound of ground turmeric, one and a half ounces of gamboge, three and a half pounds of powdered gum sandarach, three quarters of a pound of shellac, and two gallons of spirits of wine. When shaken, dissolved, and strained, add one pint of turpentine varnish, well mixed.

RED SPIRIT LACKER.

Made exactly as the gold lacker with these ingredients : two gallons of spirits of wine, one pound of dragon's blood, three pounds of Spanish annotto, three and a quarter pounds of gum sandarach, and two pints of turpentine.

PALE BRASS LACKER.

Two gallons of spirits of wine, one pound of fine pale shellac, three ounces of Cape aloes, cut small ; one ounce of gamboge, cut small.

LACKER FOR TIN.

Any good lacker laid upon tin gives it the appearance of copper or brass. It is made by coloring lac-varnish with turmeric to impart the color of brass to it, and with annotto, to give it the color of copper. If a tin plate is dipped into molten brass, the latter metal will adhere to it in a coat.

LACKER VARNISH.

A good lacker is made by coloring lac-varnish with turmeric and annotto. Add as much of these two coloring substances to the varnish as will give it the proper color ; then squeeze the varnish through a cotton cloth, when it forms lacker.

DEEP GOLD COLORED LACKER.

Seed-lac three ounces, turmeric one ounce, dragon's blood onefourth ounce, alcohol one pint ; digest for a week, frequently shaking, decant and filter.

Lackers are used upon polished metals and wood to impart the appearance of gold. If yellow is required, use turmeric, aloes, saffron, or gamboge ; for red, use annotto, or dragon's blood, to color. Turmeric, gamboge, and dragon's blood, generally afford a sufficient range of colors.

LACKERS FOR PICTURES, METAL, WOOD OR LEATHER.

1. Seed-lac eight ounces, alcohol one quart ; digest in a close vessel in a warm situation for three or four days, then decant and strain. 2. Substitute lac bleached by chlorine for seed-lac. Both are very tough, hard, and durable ; the last almost colorless.

7

MISCELLANEOUS CEMENTS.

ARMENIAN OR DIAMOND CEMENT.

This article, so much esteemed for uniting pieces of broken glass, for repairing precious stones, and for cementing them to watch cases and other ornaments, is made by soaking isinglass in water until it becomes quite soft, and then mixing it with spirit in which a little gum mastic and ammoniacum have been dissolved.

The jewellers of Turkey, who are mostly Armenians, have a singular method of ornamenting watch cases, &c., with diamonds and other precious stones, by simply glueing or cementing them on. The stone is set in silver or gold, and the lower part of the metal made flat, or to correspond with the part to which it is to be fixed ; it is then warmed gently, and has the glue applied, which is so very strong that the parts so cemented never separate ; this glue, which will strongly unite bits of glass, and even polished steel, and may be applied to a variety of useful purposes, is thus made in Turkey :—Disso ve five or six bits of gum mastic, each the size of a large pea, in as mluch spirits of wine as will suffice to render it liquid ; and in another vessel, dissolve as much isinglass, previously a little softened in water, (though none of the water must be used,) in French brandy or good rum, as will make a two-ounce vial of very strong glue, adding two small bits of gum albanum, or ammoniacum, which must be rubbed or ground till they are dissolved. Then mix the whole with a sufficient heat. Keep the glue in a vial closely stopped, and when it is to be used, set the vial in boiling water. Some persons have sold a composition under the name of Armenian cement, in England ; but this composition is badly made ; it is much too thin, and the quantity of mastic is much too small.

The following are good proportions: isinglass, soaked in water and dissolved in spirit, two ounces, (thick); dissolve in this ten grains of very pale gum ammoniac, (in tears,) by rubbing them together ; then add six large tears of gum mastic, dissolved in the least possible quantity of rectified spirit.

Isinglass, dissolved in proof spirit, as above, three ounces ; bottoms of mastic varnish (thick but clear) one and a half ounces ; mix well.

When carefully made, this cement resists moisture, and dries colorless. As usually met with, it is not only of very bad quality, but sold at exorbitant prices.

[CEMENTS FOR MENDING EARTHERN AND GLASS WARE.

1. Heat the article to be mended, a little above boiling water heat, then apply a thin coating of gum shellac, on both surfaces of the broken vessel, and when cold it will be as strong as it was originally. 2. Dissolve gum shellac in alcohol, apply the solution, and bind the parts firmly together until the cement is perfectly dry.

CEMENT FOR STONEWARE.

Another cement in which an analogous substance, the curd or caseum of milk is employed, is made by boiling slices of skim-milk cheese into a glucy consistence in a great quantity of water, and then incorporating it with quicklime on a slab with a muller, or in a marble mortar. When this compound is applied warm to broken edges of stoneware, it unites them very firmly after it is cold.

IRON-RUST CEMENT.

The iron-rust cement is made of from fifty to one hundred parts of iron borings, pounded and sifted, mixed with one part of sal-ammoniac, and when it is to be applied moistened with as much water as will give it a pasty consistency. Formerly flowers of sulphur were used, and much more sal-ammoniac in making this cement, but with decided disadvantage, as the union is effected by oxidizement, consequent expansion and solidification of the iron powder, and any heterogeneous matter obstructs the effect. The best proportion of sal-ammoniac is, I believe, one per cent of the iron borings. Another composition of the same kind is made by mixing four parts of fine borings or filings of iron, two parts of potter's clay, and one part of pounded potsherds, and making them into a paste with salt and water. When this cement is allowed to concrete slowly on iron joints, it becomes very hard.

FOR MAKING ARCHITECTURAL ORNAMENTS IN RELIEF.

For making architectural ornaments in relief, a moulding composition is formed of chalk, glue, and paper paste. Even statues have been made with it, the paper aiding the cohesion of the mass.

Mastics of a resinous or bituminous nature, which must be softened or fused by heat, are the following :—

VARLEY'S MASTIC.

Mr. S. Varley's consists of sixteen parts of whiting sifted and thoroughly dried by a red heat, adding when cold a melted mixture of sixteen parts of black rosin and one of bees'-wax, and stirring well during the cooling.

ELECTRICAL AND CHEMICAL APPARATUS CEMENT.

Electrical and chemical apparatus cement consists of 5 lbs. of rosin, 1 of bees'-wax, 1 of red ochre, and two table-spoonsful of Paris plaster, all melted together. A cheaper one for cementing voltaic plates into wooden troughs is made with 6 pounds of rosin, 1 pound of red ochre, ½ of a pound of plaster of Paris, and ¼ of a pound of linseed oil. The ochre and the plaster of Paris should be calcined beforehand, and added to the other ingredients in their melted state. The thinner the stratum of cement that is interposed, the stronger, generally speaking, is the junction.

CEMENT FOR IRON TUBES, BOILERS, ETC.

Finely powdered iron sixty-six parts, sal-ammoniac one part, water a sufficient quantity to form into paste.

CEMENT FOR IVORY, MOTHER OF PEARL, ETC.

Dissolve one part of isinglass and two of white glue in thirty of water, strain and evaporate to six parts. Add one-thirtieth part of gum mastic, dissolved in half a part· of alcohol, and one part of white zinc. When required for use, warm and shake up.

CEMENT FOR HOLES IN CASTINGS.

The best cement for this purpose is made by mixing one part of sulphur in powder, two parts of sal-ammoniac, and eighty parts of clean powdered iron turnings. Sufficient water must be added to make it into a thick paste, which should be pressed into the holes or seams which are to be filled up. The ingredients composing this cement should be kept separate, and not mixed until required for use. It is to be applied cold, and the casting should not be used for two or three days afterwards.

CEMENT FOR COPPERSMITHS AND ENGINEERS.

Boiled linseed oil and red lead mixed together into a putty are often used by coppersmiths and engineers, to secure joints. The washers of leather or cloth are smeared with this mixture in a pasty state.

A CHEAP CEMENT.

Melted brimstone, either alone, or mixed with rosin and brick dust, forms a tolerably good and very cheap cement.

PLUMBER'S CEMENT.

Plumber's cement consists of black rosin one part, brick dust two parts, well incorporated by a melting heat.

CEMENT FOR BOTTLE-CORKS.

The bituminous or black cement for bottle-corks consists of pitch hardened by the addition of rosin and brick-dust.

CHINA CEMENT.

Take the curd of milk, dried and powdered, ten ounces ; quicklime one ounce.; camphor two drachms. Mix, and keep in closely stopped bottles. When used, a portion is to be mixed with a little water into a paste, to be applied quickly.

CEMENT FOR LEATHER.

A mixture of India-rubber and shell-lac varnish makes a very adhesive leather cement. A strong solution of common isinglass, with a little diluted alcohol added to ⬤, makes an excellent cement for leather.

MARBLE CEMENT.

Take plaster of paris, and soak it in a saturated solution of alum, then bake the two in an oven, the same as gypsum is baked to make it plaster of paris ; after which they are ground to powder. It is then used as wanted, being mixed up with water like plaster and applied. It sets into a very hard composition capable of taking a very high polish. It may be mixed with various coloring minerals to produce a cement of any color capable of imitating marble.

A GOOD CEMENT.

Shellac dissolved in alcohol, or in a solution of borax, forms a pretty good cement.

CEMENT FOR MARBLE WORKERS AND COPPERSMITHS.

White of egg alone, or mixed with finely sifted quicklime, will answer for uniting objects which are not exposed to moisture. The latter combination is very strong, and is much employed for joining pieces of spar and marble ornaments. A similar composition is used by coppersmiths to secure the edges and rivets of boilers ; only bullock's blood is the albuminous matter used instead of white of egg.

TRANSPARENT CEMENT FOR GLASS.

Dissolve one part of India-rubber in 64 of chloroform, then add gum mastic in powder 16 to 24 parts, and digest for two days with frequent shaking. Apply with a camels-hair brush.

CEMENT TO MEND IRON POTS AND PANS.

Take two parts of sulphur, and one part, by weight, of fine black lead ; put the sulphur in an old iron pan, holding it over the fire until it begins to melt, then add the lead ; stir well until all is mixed and melted ; then pour out on an iron plate, or smooth stone. When cool, break into small pieces. A sufficient quantity of this compound being placed upon the crack of the iron pot to be mended, can be soldered by a hot iron in the same way a tinsmith solders his sheets. If there is a small hole in the pot, drive a copper rivet in it and then solder over it with this cement.

CEMENT TO RENDER CISTERNS AND CASKS WATER TIGHT.

An excellent cement for resisting moisture is made by incorporating thoroughly eight parts of melted glue, of the consistence used by carpenters, with four parts of linseed oil, boiled into varnish with litharge. This cement hardens in about forty-eight hours, and renders the joints of wooden cisterns and casks air and water tight. A compound of glue with one-fourth its weight of Venice turpentine, made as above, serves to cement glass, metal and wood, to one another Fresh-made cheese curd, and old skim-milk cheese, boiled in water to a slimy consistence, dissolved in a solution of bicarbonate of potash

7*

are said to form a good cement for glass and porcelain. The gluten of wheat, well prepared, is also a good cement. White of eggs, with flour and water well-mixed, and smeared over linen cloth, forms a ready lute for steam joints in small apparatus.

CEMENT FOR REPAIRING FRACTURED BODIES OF ALL KINDS.

White lead ground upon a slab with linseed oil varnish, and kept out of contact of air, affords a cement capable of repairing fractured bodies of all kinds. It requires a few weeks to harden. When stone or iron are to be cemented together, a compound of equal parts of sulphur with pitch answers very well.

CEMENTS FOR CRACKS IN WOOD.

Make a paste of slacked lime one part, rye-meal two parts, with a sufficient quantity of linseed oil. Or, dissolve one part of glue in sixteen parts of water, and when almost cool stir in sawdust and prepared chalk a sufficient quantity. Or, oil-varnish thickened with a mixture of equal parts of white-lead, red-lead, litharge, and chalk.

CEMENT FOR JOINING METALS AND WOOD.

Melt rosin and stir in calcined plaster until reduced to a paste, to which add boiled oil a sufficient quantity, to bring it to the consistence of honey ; apply warm. Or, melt rosin 180 parts, and stir in burnt umber 30, calcined plaster 15, and boiled oil 8 parts.

GAS FITTERS' CEMENT.

Mix together, resin four and one-half parts, wax one part, and venetian red three parts.

IMPERVIOUS CEMENT FOR APPARATUS, CORKS, ETC.

Zinc-white rubbed up with copal varnish to fill up the indentures; when dry, to be covered with the same mass, somewhat thinner, and lastly with copal varnish alone.

CEMENT FOR FASTENING BRASS TO GLASS VESSELS.

Melt rosin 150 parts, wax 30, and add burnt ochre 30, and calcined plaster 2 parts. Apply warm.

CEMENT FOR FASTENING BLADES, FILES, ETC.

Shellac two parts, prepared chalk one, powdered and mixed. The opening for the blade is filled with this powder, the lower end of the iron heated and pressed in.

HYDRAULIC CEMENT PAINT.

If hydraulic cement be mixed with oil, it forms a first-rate anti-combustible and excellent water-proof paint for roofs of buildings, outhouses, walls, &c.

BUILDERS' CEMENTS.

CEMENT FOR TERRACES, FLOORS, ROOFS, RESERVOIRS, ETC.

In certain localities where a limestone impregnated with bitumen occurs, it is dried, ground, sifted, and then mixed with about its own weight of melted pitch, either mineral, vegetable, or that of cold tar. When this mixture is getting semifluid, it may be moulded into large slabs or tiles in wooden frames lined with sheet iron, previously smeared over with common lime mortar, in order to prevent adhesion to the moulds, which, being in moveable pieces, are easily dismounted so as to turn out the cake of artificial bituminous stone. This cement is manufactured upon a great scale in many places, and used for making Italian terraces, covering the floors of balconies, flat roofs, water reservoirs, water conduits, &c. When laid down, the joints must be well run together with hot irons. The floor of the terrace should be previously covered with a layer of Paris plaster or common mortar, nearly an inch thick, with a regular slope of one inch to the yard. Such bituminous cement weighs 144 pounds the cubic foot; or a foot of square surface, one inch thick, weighs 12 pounds. Sometimes a second layer of these slabs or tiles is applied over the first, with the precaution of making the seams or joints of the upper correspond with the middle of the under ones. Occasionally a bottom bed, of coarse cloth or gray paper, is applied. The larger the slabs are made, as far as they can be conveniently tsansported and laid down, so much the better.

MASTIC CEMENT FOR COVERING THE FRONTS OF HOUSES.

Fifty parts, by measure, of clean dry sand, fifty of limestone (not burned) reduced to grains like sand, or marble dust, and 10 parts of red lead, mixed with as much boiled linseed oil, as will make it slightly moist. The brick, to receive it, should be covered with three coats of boiled oil, laid on with a brush, and suffered to dry, before the mastic is put on. It is laid on with a trowel like plaster, but it is not so moist. It becomes hard as stone in a few months. Care must be exercised not to use too much oil.

CEMENT FOR OUTSIDE BRICK WALLS.

Cement for the outside of brick walls, to imitate stone, is made of clean sand 90 parts, litharge 5 parts, plaster of Paris 5 parts, moistened with boiled linseed oil. The bricks should receive two or three coats of oil before the cement is applied.

CEMENT FOR COATING THE FRONTS OF BUILDINGS.

The cement of dihl for coating the fronts of buildings consists of linseed oil, rendered dry by boiling with litharge, and mixed with porcelain clay in fine powder, to give it the consistence of stiff mortar.

Pipe-clay would answer equally well if well dried, and any color might be given with ground bricks, or pottery. A little oil of turpentine to thin this cement aids its cohesion upon stone, brick or wood. It has been applied to sheets of wire cloth, and in this state laid upon terraces, in order to make them water tight ; but it is a little less expensive than lead.

CEMENT FOR STEPS AND BRICK WALLS.

A cement which gradually indurates to a stony consistence, may be made by mixing twenty parts of clean river sand, two of litharge, and one of quicklime, into a thin putty with linseed oil. The quicklime may be replaced with litharge. When this cement is applied to mend broken pieces of stone, as steps of stairs, it acquires after some time a stony hardness. A similar composition has been applied to coat over brick walls, under the name of mastic.

A HARD CEMENT FOR SEAMS.

An excellent cement for seams in the roofs of houses, or for any other exposed places, is made with white lead, dry white sand, and as much oil as will make it into the consistency of putty. This cement gets as hard as stone in a few weeks. It is a good cement for filling up cracks in exposed parts of brick buildings ; and for pointing up the base of chimneys, where they project through the roofs of shingled houses.

ANOTHER GOOD CEMENT.

Dissolve one pound of alum in boiling water, and while it is boiling add five pounds of brown soap, cut into small pieces ; boil the mixture about fifteen minutes. It then becomes sticky like shoemaker's wax. Now mix it with whiting to a proper consistency for filling up seams, &c. It becomes partially hard after a few months, and strongly adheres to wood. The wood should be perfectly dry. To make it adhere it must be well pressed down. When dry it is impervious to water, and is slightly elastic.

CEMENT FOR TILE-ROOFS,

The best cement for closing up seams in tile-roofs is composed of equal parts of whiting and dry sand and 25 per cent of litharge, made into the consistency of putty with linseed oil. It is not liable to crack when cold, nor melt, like coal-tar and asphalt, with the heat of the sun.

COARSE STUFF.

Coarse stuff, or lime and hair, as it is sometimes called, is prepared in the same way as common mortar, with the addition of hair procured from the tanner, which must be well mixed with the mortar by means of a three-pronged rake, until the hair is equally distributed throughout the composition. The mortar should be first formed, and when the lime and sand have been thoroughly mixed, the hair should be added by degrees, and the whole so thoroughly united, that the hair shall appear to be equally distributed throughout.

PARKER'S CEMENT.

This cement, which is perhaps the best of all others for stucco, as it is not subject to crack or flake off, is now very commonly used, and is formed by burning argillaceous clay in the same manner that lime is made. It is then reduced to powder. The cement, as used by the plasterer, is sometimes employed alone, and sometimes it is mixed with sharp sand ; and it has then the appearance, and almost the strength, of stone. As it is impervious to water, it is very proper for lining tanks and cisterns.

HAMELEIN'S CEMENT.

This cement consists of earthy and other substances insoluble in water, or nearly so ; and these may be either those which are in their natural state, or have been manufactured, such as earthenware and china ; those being always preferred which are least soluble in water, and have the least color. When these are pulverized, some oxide of lead is added, such as litharge, gray oxide, or minium, reduced to a fine powder ; and to the compound is added a quantity of pulverized glass or flint stones, the whole being thoroughly mixed and made into a proper consistence with some vegetable oil, as that of linseed. This makes a durable stucco or plaster, that is impervious to wet, and has the appearance of stone.

The proportion of the several ingredients is as follows : — to every five hundred and sixty pounds of earth, or earths, such as pit sand, river sand, rock sand, pulverized earthenware or porcelain, add forty pounds of litharge, two pounds of pulverized glass or flint, one pound of minium, and two pounds of gray oxide of lead. Mix the whole together, and sift it through sieves of different degrees of fineness, according to the purposes to which the cement is to be applied.

The following is the method of using it : — To every thirty pounds weight of the cement in powder, add about one quart of oil, either linseed, walnut, or some other vegetable oil, and mix it in the same manner as any other mortar, pressing it gently together, either by treading on it, or with the trowel ; it has then the appearance of moistened sand. Care must also be taken that no more is mixed at one time than is required for use, as it soon hardens into a solid mass. Before the cement is applied, the face of the wall to be plastered should be brushed over with oil, particularly if it be applied to brick, or any other substance that quickly imbibes the oil ; if to wood, lead, or any substance of a similar nature, less oil may be used.

PLASTER IN IMITATION OF MARBLE—SCAGLIOLA.

This species of work is exquisitely beautiful when done with taste and judgment, and is so like marble to the touch, as well as appearance, that it is scarcely possible to distinguish the one from the other. We shall endeavor to explain its composition, and the manner in which it is applied ; but so much depends upon the workman's execution, that it is impossible for any one to succeed in an attempt to work with it without some practical experience.

Procure some of the purest gypsum, and calcine it until the large masses have lost the brilliant, sparkling appearance by which they are characterized, and the whole mass appears uniformly opaque. This calcined gypsum is reduced to powder, and passed through a very fine sieve, and mixed up, as it is wanted for use, with glue, isinglass, or some other material of the same kind. This solution is colored with the tint required for the scagliola ; but when a marble of various colors is to be imitated, the several colored compositions required by the artist must be placed in separate vessels, and they are then mingled together in nearly the same manner that the painter mixes his color on the pallet. Having the wall or column prepared with rough plaster, it is covered with the composition, and the colors intended to imitate the marble, of whatever kind it may be, are applied when the floating is going on.

It now only remains to polish the work, which, as soon as the composition is hard enough, is done by rubbing it with pumice-stone, the work being kept wet with water applied by a sponge. It, is then polished with Tripoli and charcoal, with a piece of fine linen, and finished with a piece of felt, dipped in a mixture of oil and Tripoli, and afterwards with pure oil.

MALTHA, OR GREEK MASTIC.

This is made by mixing lime and sand in the manner of mortar, and making it into a proper consistency with milk or size, instead of water.

FINE STUFF.

This is made by slaking lime with a small portion of water, after which so much water is added as to give it the consistence of cream. It is then allowed to settle for some time, and the superfluous water is poured off, and the sediment is suffered to remain till evaporation reduces it to a proper thickness for use. For some kinds of work, it is necessary to add a small portion of hair.

STUCCO FOR INSIDE OF WALLS.

This stucco consists of fine stuff already described, and a portion of fine washed sand, in the proportion of one of sand to three of fine stuff. Those parts of interior walls are finished with this stucco which are intended to be painted. In using this material, great care must be taken that the surface be perfectly level, and to secure this it must be well worked with a floating tool or wooden trowel. This is done by sprinkling a little water occasionally on the stucco, and rubbing it in a circular direction with the float, till the surface has attained a high gloss. The durability of the work very much depends upon the care with which this process is done ; for if it be not thoroughly worked, it is apt to crack.

HIGGINS' STUCCO.

To fifteen pounds of the best stone lime, add fourteen pounds of bone ashes, finely powdered, and about ninety-five pounds of clean, washed sand, quite dry, either coarse or fine, according 'to the nature of the work in hand. These ingredients must be intimately mixed, and kept from the air till wanted. When required for use, it must be mixed up into a proper consistence for working with lime water, and used as speedily as possible.

GAUGE STUFF.

This is chiefly used for mouldings and cornices which are run or formed with a wooden mould. It consists of about one-fifth of plaster of Paris, mixed gradually with four-fifths of fine stuff. When the work is required to set very expeditiously, the proportion of plaster of Paris is increased. It is often necessary that the plaster to be used should have the property of setting immediately it is laid on, and in all such cases gauge stuff is used, and consequently it is extensively employed for cementing ornaments to walls or ceilings, as well as for casting the ornaments themselves.

COMPOSITION.

This is frequently used, instead of plaster of Paris, for the ornamental parts of buildings, as it is more durable, and becomes in time as hard as stone itself. It is of great use in the execution of the decorative parts of architecture, and also in the finishings of picture frames, being a cheaper method than carving by nearly eighty per cent.

It is made as follows : — Two pounds of the best whitening, one pound of glue, and half a pound of linseed oil are heated together, the composition being continually stirred until the different substances are thoroughly incorporated. Let the compound cool, and then lay it on a stone covered with powdered whitening, and heat it well until it becomes of a tough and firm consistence. It may then be put by for use, covered with wet cloths to keep it fresh. When wanted for use, it must be cut into pieces, adapted to the size of the mould, into which it is forced by a screw press. The ornament, or cornice, is fixed to the frame or wall with glue or with white lead.

FOUNDATIONS OF BUILDINGS.

The nature and condition of the soil upon which houses are to be built should receive far more attention than is usually bestowed upon such subjects. A soil which is spongy and damp, or contains much loose organic matter, is generally unhealthy ; whereas a dry, porus soil affords a healthy site for buildings. Wherever we find a soil deficient in gravel or sand, or where gravel and sand-beds are underlaid with clay, there should be a thorough sub-soil drainage, because the clay retains the water, and a house built in such a spot would otherwise always be damp and unhealthy.

When the sub-soil is swampy, which is the case with many portions
of various cities that have been filled in with what is called *made
earth*, fever is liable to prevail in houses built in such localities,
owing to the decay of organic matter underneath, and its ascension
in the form of gas through the soil. When good drainage cannot be
effected in such situations, and it is found necessary to build houses
on them, they should all have solid floors of concrete, laid from the
outside of the foundations and covering the whole area over which
the structure is erected. These floors tend to prevent dampness in
houses, consequently they are more comfortable and healthy than
they otherwise would be. Such floors also tend to prevent the cracking of the walls, owing to the solidity and firmness imparted to their
foundations.

CONCRETE FLOORS.

The lower floors of all the cellars of houses should be composed of a
bed of concrete about three inches thick. This would tend to render
them dry, and more healthy, and at the same time prevent rats from
burrowing under the walls from the outside, and coming up under
the floor—the method pursued by these vermin where houses are
erected on a sandy soil. This concrete should be made of washed
gravel and hydraulic cement. Common mortar mixed with pounded
brick and washed gravel, makes a concrete for floors nearly as good
as that formed with hydraulic cement. Such floors become very hard,
and are much cheaper than those of brick or flagstones.

FIRE-PROOF COMPOSITION TO RESIST FIRE FOR FIVE HOURS.

Dissolve, in cold water, as much pearlash as it is capable of holding
in solution, and wash or daub with it all the boards, wainscoting,
timber, &c. Then diluting the same liquid with a little water, add to
it such a portion of fine yellow clay as will make the mixture the same
consistence as common paint ; stir in a small quantity of paperhang-
er's flour paste to combine both the other substances. Give three
coats of this mixture. When *dry*, apply the following mixture :—
Put into a pot equal quantities of finely pulverized iron filings, brick
dust, and ashes : pour over them size or glue water ; set the whole
near a fire, and when warm stir them well together. With this liquid
composition, or size, give the wood one coat ; and on its getting dry,
give it a second coat. It resists fire for five hours, and prevents the
wood from ever bursting into flames. It resists the ravages of fire,
so as only to be reduced to coal or embers, without spreading the con-
flagration by additional flames ; by which five clear hours are gained
in removing valuable effects to a place of safety, as well as rescuing
the lives of all the family from danger ! Furniture, chairs, tables,
&c., particularly staircases, may be so protected. Twenty pounds of
finely sifted yellow clay, a pound and a half of flour for making the
paste, and one pound of pearlash, are sufficient to prepare a square
rood of deal-boards.

MISCELLANEOUS RECEIPTS.

TO POLISH WAINSCOT AND MAHOGANY.

A very good polish for wainscot may be made in the following manner : Take as much beeswax as required, and, placing it in a glazed earthen pan, add as much spirits of wine as will cover it, and let it dissolve without heat. Add either one ingredient as is required, to reduce it to the consistence of butter. When this mixture is well rubbed into the grain of the wood, and cleaned off with clean linen, it gives a good gloss to the work.

IMITATION OF MAHOGANY.

Plane the surface smooth, and rub with a solution of nitrous acid. Then apply with a soft brush one ounce of dragon's blood, dissolved in about a pint of alcohol, and with a third of an ounce of carbonate of soda, mixed and filtered. When the brilliancy of the polish diminishes, it may be restored by the use of a little cold drawn linseed oil.

FURNITURE VARNISH.

White wax six ounces, oil of turpentine one pint ; dissolve by a gentle heat. *Used* to polish wood by friction.

TO MAKE GLASS PAPER.

Take any quantity of broken glass (that with a greenish hue is the best), and pound it in an iron mortar. Then take several sheets of paper, and cover them evenly with a thin coat of glue, and, holding them to the fire, or placing them upon a hot piece of wood or plate of iron, sift the pounded glass over them. Let the several sheets remain till the glue is set, and shake off the superfluous powder, which will do again. Then hang up the papers to dry and harden. Paper made in this manner is much superior to that generally purchased at the shops, which chiefly consists of fine sand. To obtain different degrees of fineness, sieves of different degrees of fineness must be used. Use thick paper.

TO MAKE STONE PAPER.

As, in cleaning wood-work, particularly deal and other soft woods, one process is sometimes found to answer better than another, we may describe the manner of manufacturing a stone paper, which, in some cases, will be preferred to sand paper, as it produces a good face, and is less liable to scratch the work. Having prepared the paper as already described, take any quantity of powdered pumice-stone, and sift it over the paper through a sieve of moderate fineness. When the surface has hardened, repeat the process till a tolerably thick coat has been formed upon the paper, which, when dry, will be fit for use.

8

WHITEWASH.

The best method of making a whitewash for outside exposure is to slack half a bushel of lime in a barrel, add one pound of common salt, half a pound of the sulphate of zinc, and a gallon of sweet milk.

PAINT FOR COATING WIRE WORK.

Boil good linseed oil with as much litharge as will make it of the consistency to be laid on with the brush ; add lampblack at the rate of one part to every ten, by weight of the litharge ; boil three hours over a gentle fire. The first coat should be thinner than the following coats.

TO BLEACH SPONGE.

Soak it well in dilute muriatic acid for twelve hours. Wash well with water, to remove the lime, then immerse it in a solution of hyposulphite of soda, to which dilute muriatic acid has been added a moment before. After it is bleached sufficiently remove it, wash again, and dry it. It may thus be bleached almost snow white.

LAC VARNISH FOR VINES.

Grape vines may be pruned at any period without danger from loss of bleeding, by simply covering the cut parts with varnish made by dissolving stick-lac in alcohol. The lac varnish soon dries, and forms an impenetrable coat to rain ; it may also be applied with advantage in coating the wounds of young trees.

RAZOR PASTE.

1. Levigated oxide of tin (prepared putty powder) 1 oz. ; powdered oxalic acid 1-4 oz. ; powdered gum 20 grs. ; make it into a stiff paste with water, and evenly and thinly spread it over the strop. With very little friction, this paste gives a fine edge to the razor, and its efficiency is still further increased by moistening it.

2. Emery reduced to an impalpable powder 2 parts ; spermaceti ointment 1 part ; mix together, and rub it over the strop.

3. Jewellers' rouge, blacklead, and suet, equal parts ; mix.

LEATHER VARNISH.

Durable leather varnish is composed of boiled linseed oil, in which a drier, such as litharge, has been boiled. It is colored with lampblack. This varnish is used for making enamelled leather. Common leather varnish, which is used as a substitute for blacking, is made of thin lac-varnish colored with ivory black.

TO KEEP TIRES TIGHT ON WHEELS.

Before putting on the tires fill the felloes with linseed oil, which is done by heating the oil in a trough to a boiling heat, and keeping the wheel, with a stick through the hub, in the oil, for an hour The wheel is turned round until every felloe is kept in the oil one hour.

CUTTING GLASS.

To cut bottles, shades, or other glass vessels neatly, heat a rod of iron to redness, and having filled your vessel the exact height you wish it to be cut, with oil of any kind, you proceed very gradually to dip the red hot iron into the oil, which, heating all along the surface, suddenly the glass chips and cracks right round, when you can lift off the upper portion clean by the surface of the oil.

PREPARED LIQUID GLUE.

Take of best white glue 16 ounces; white lead, dry, 4 ounces; rain water 2 pints; alcohol 4 ounces. With constant stirring dissolve the glue and lead in the water by means of a water-bath. Add the alcohol, and continue the heat for a few minutes. Lastly pour into bottles while it is still hot.

LIQUID GLUES.

Dissolve 33 parts of best (Buffalo) glue on the steam bath in a porcelain vessel, in 36 parts of water. Then add gradually, stirring constantly, 3 parts of aqua fortis, or as much as is sufficient to prevent the glue from hardening when cool. Or, dissolve one part of powdered alum in 120 of water, add 120 parts of glue, 10 of acetic acid and 40 of alcohol, and digest.

MARINE GLUE.

Dissolve 4 parts of india rubber in 34 parts of coal tar naphtha—aiding the solution with heat and agitation, add to it 64 parts of powdered shellac, which must be heated in the mixture, till the whole is dissolved. While the mixture is hot it is poured upon metal plates in sheets like leather. When required for use, it is heated in a pot, till soft, and then applied with a brush to the surfaces to be joined. Two pieces of wood joined with this glue can scarcely be sundered.

AN EXCELLENT PASTE FOR ENVELOPES.

Mix in equal quantities gum-arabic (substitute dextrine) and water in a phial, place it near a stove, or on a furnace register, and stir or shake it well, until it dissolves. Add a little alcohol to prevent its souring.

DEXTRINE, OR BRITISH GUM.

Dry potato-starch heated from 300° to 600° until it becomes brown, soluble in cold water, and ceases to turn blue with iodine. *Used* by calico printers and others, instead of gum arabic.

GUM MUCILAGE.

A little oil of cloves poured into a bottle containing gum mucilage prevents the latter from becoming sour and putrid; this essential oil possesses great antiseptic powers.

FLOUR PASTE.

Too numerous to mention are the little conveniences of having a little flour paste always at hand, as those made of any of the gums impart a glaze to printed matter, and make it rather difficult to read. Dissolve a tablespoonful of alum in a quart of warm water, and when cold, stir in as much flour as will give it the consistency of thick cream, being particular to beat up all the lumps, then stir in as much powdered resin as will stand on a dime, then throw in half a dozen cloves, merely to give a pleasant odor. Next, have a vessel on the fire which has a teacupful or more of boiling water, pour the flour mixture on the *boiling* water, stir it well all the time ; in a very few minutes it will be of the consistence of mush ; pour it out in an earthen or china vessel ; let it cool ; lay a cover on it, and put it in a cool place. It will keep for months. When needed for use, take out a portion and soften it with warm water. Keep it covered an inch or two in water to prevent the surface from drying up.

SEALING-WAX FOR FRUIT-CANS.

Beeswax, ½ oz. ; English vermillion, 1½ ozs. ; gum shellac, 2½ ozs. ; rosin, 8 ozs. Take some cheap iron vessel that you can always keep for the purpose, and put in the rosin and melt it, and stir in the vermillion. Then add the shellac, slowly, and stir that in, and afterward the beeswax. When wanted for use at any after time, set it upon a slow fire and melt it so you can dip bottle-nozzles in. For any purpose, such as an application to trees, where you want it tougher than the above preparation will make it, add a little more beeswax, and leave out the vermillion.

If the vermillion is left out in the above, the wax will be all the better for it, as it is merely used for coloring purposes.

FUSIBLE METAL.

1. Bismuth 8 parts ; lead 5 parts ; tin 3 parts ; melt together, Melts below 212 degrees Fahr. 2. Bismuth 2 parts ; lead 5 parts ; tin 3 parts. Melts in boiling water. 3. Lead 3 parts; tin 2 parts; bismuth 5 parts ; mix. Melts at 197 deg. Fahr.

Remarks. The above are used to make toy-spoons, to surprise children by their melting in hot liquors ; and to form pencils for writing on asses' skin, or paper prepared by rubbing burnt hartshorn into it.

METALLIC CEMENT.

M. Greshiem states that an alloy of copper and mercury, prepared as follows, is capable of attaching itself firmly to the surfaces of metal, glass, and porcelain. From twenty to thirty parts of finely divided copper, obtained by the reduction of oxide of copper with hydrogen, or by precipitation from solution of its sulphate with zinc, are made into a paste with oil of vitrol and seventy parts of mercury added, the whole being well triturated. When the amalgamation is complete, the acid is removed by washing with boiling water, and the compound allowed to cool. In ten or twelve hours, it becomes sufficiently hard to receive a brilliant polish, and to scratch the surface of tin or gold. By heat it assumes the consistence of wax ; and, as it does not contract on cooling, M. Greshiem recommends its use by dentists for stopping teeth.

ARTIFICIAL GOLD.

This is a new metallic alloy which is now very extensively used in France as a substitute for gold. Pure copper 100 parts, zinc, or preferably tin 17 parts, magnesia 6 parts, sal ammoniac 3·6 parts, quick lime 1·8 parts, tartar of commerce 9 parts, are mixed as follows : The copper is first melted, then the magnesia, sal ammoniac, lime, and tartar, are then added, separately and by degrees, in the form of powder ; the whole is now briskly stirred for about half an hour, so as to mix thoroughly ; and then the zinc is added in small grains by throwing it on the surface and stirring till it is entirely fused ; the crucible is then covered and the fusion maintained for about 35 minutes. The surface is then skimmed and the alloy is ready for casting.

It has a fine grain, is malleable and takes a splendid polish. It does not corrode readily, and for many purposes is an excellent substitute for gold. When tarnished, its brilliancy can be restored by a little acidulated water. If tin be employed instead of zinc the alloy will be more brilliant. It is very much used in France, and must ultimately attain equal popularity here.

OR-MOLU.

The or-molu of the brass founder, popularly known as an imitation of red gold, is extensively used by the French workmen in metals. It is generally found in combination with grate and stove work. It is composed of a greater portion of copper and less zinc than ordinary brass, is cleaned readily by means of acid, and is burnished with facility. To give this material the rich appearance, it is not unfrequently brightened up after "dipping" (that is cleaning in acid) by means of a scratch brush (a brush made of fine brass wire), the action of which helps to produce a very brilliant gold-like surface. It is protected from tarnish by the application of lacker. •

BLANCHED COPPER.

Fuse 8 ounces of copper and ½ ounce of neutral arsenical salt, with a flux made of calcined borax, charcoal dust and powdered glass.

BROWNING GUN BARRELS.

The tincture of iodine diluted with one-half its bulk of water, is a superior liquid for browning gun barrels.

SILVERING POWDER FOR COATING COPPER.

Nitrate of silver 30 grains, common salt 30 grains, cream of tartar 3¼ drachms ; mix, moisten with water, and apply.

8*

ALLOY FOR JOURNAL BOXES.

The best alloy for journal boxes is composed of copper, 24 lbs. ; tin, 24 lbs. ; and antimony, 8 lbs. Melt the copper first, then add the tin, and lastly the antimony. It should be first run into ingots, then melted and cast in the form required for the boxes.

ALLOY FOR BELLS OF CLOCKS.

The bells of the *pendules*, or ornamental clocks, made in Paris, are composed of copper 72.00, tin 26.56, iron 1.44, in 100 parts.

AN ALLOY FOR TOOLS.

An alloy of 1000 parts of copper and 14 of tin is said to furnish tools, which hardened and sharpened in the manner of the ancients, afford an edge nearly equal to that of steel.

ALLOY FOR CYMBALS AND GONGS.

An alloy for cymbals and gongs is made of 100 parts of copper with about 25 of tin. To give this compound the sonorous property in the highest degree, the piece should be ignited after it is cast, and then plunged immediately into cold water.

SOLDER FOR STEEL JOINTS.

Silver 19 pennyweights, copper 1 pennyweight, brass 2 pennyweights. Melt under a coat of charcoal dust.

SOFT GOLD SOLDER.

Is composed of four parts gold, one of silver, and one of copper. It can be made softer by adding brass, but the solder becomes more liable to oxidize.

FILES.

Allow dull files to lay in diluted sulphuric acid until they are bit deep enough.

TO PREVENT RUSTING.

Boiled linseed oil will keep polished tools from rusting if it is allowed to dry on them. Common sperm oil will prevent them from rusting for a short period. A coat of copal varnish is frequently applied to polished tools exposed to the weather.

ANTI-ATTRITION, AND AXLE-GREASE.

One part of fine black lead, ground perfectly smooth, with four parts of lard.

TO GALVANIZE.

Take a solution of nitro-muriate of gold (gold dissolved in a mixture of aquafortis and muriatic acid) and add to a gill of it a pint of ether or alcohol, then immerse your copper chain in it for about 16 minutes, when it will be coated with a film of gold. The copper must be perfectly clean and free from oxyd, grease, or dirt, or it will not take on the gold.

RARE AND VALUABLE COMPOSITIONS.

*Receipts for the use of Mechanists, Iron and Brass Founders,
Tinmen, Coppersmiths, Turners, Dentists, Finishers of Brass,
Britannia, and German Silver, and for other useful and im-
portant purposes in the Practical Arts.*

The larger number of the following Receipts are the result of
inquiries and experiments by a practical operative. Most of those
which relate to the mixing of metals and to the finishing of manufactured articles, have been thoroughly tested by him, and will be found
to produce the results desired and expected. The others have been
collected from eminent scientific works.

No. 1. YELLOW BRASS, *for Turning.—(Common article.)*—Copper,
20 lbs.; Zinc, 10 lbs.; Lead from 1 to 5 ozs.
Put in the Lead last before pouring off.

No. 2. RED BRASS, *for Turning.* — Copper, 24 lbs.; Zinc, 5 lbs.;
Lead, 8 ozs.
Put in the Lead last before pouring off.

No. 3. RED BRASS, *free, for Turning.* — Copper, 160 lbs.; Zinc, 50
lbs.; Lead, 10 lbs.; Antimony, 4½ ozs.

No. 4. ANOTHER BRASS, *for Turning.*—Copper, 32 lbs.; Zinc, 10
lbs.; Lead, 1 lb.

No. 5. BEST RED BRASS, *for Fine Castings.*—Copper, 24 lbs.;
Zinc, 5 lbs.; Bismuth, 1 oz.
Put in the Bismuth last before pouring off.

No. 6. BRONZE METAL. — Copper, 7 lbs.; Zinc, 3 lbs.; Tin, 2 lbs.

No. 7. BRONZE METAL. — Copper, 1 lb.; Zinc, 12 lbs.; Tin, 8 lbs.

No. 8. BELL METAL, *for large Bells.* — Copper, 100 lbs.; Tin, from
20 to 25 lbs.

No. 9. BELL METAL, *for small Bells.* — Copper, 3 lbs.; Tin, 1 lb.

No. 10. COCK METAL.— Copper, 20 lbs.; Lead, 8 lbs.; Litharge, 1 oz.;
Antimony, 3 ozs.

No. 11. HARDENING FOR BRITANNIA. — *(To be mixed separately
from the other ingredients)*— Copper, 2 lbs.; Tin, 1 lb.

No. 12. GOOD BRITANNIA METAL. — Tin, 150 lbs.; Copper, 3 lbs.;
Antimony, 10 lbs.

No. 13. BRITANNIA METAL, *2d quality.*—Tin, 140 lbs.; Copper, 3 lbs.;
Antimony, 9 lbs.

No. 14. BRITANNIA METAL, *for Casting.* — Tin, 210 lbs.; Copper,
4 lbs.; Antimony, 12 lbs.

No. 15. BRITANNIA METAL, *for Spinning.*— Tin, 100 lbs.; Britannia
Hardening, 4 lbs.; Antimony, 4 lbs.

No. 16. BRITANNIA METAL, *for Registers.* — Tin, 100 lbs.; Hardening, 8 lbs.; Antimony, 8 lbs.

No. 17. BEST BRITANNIA, *for Spouts.*— Tin, 140 lbs.; Copper, 3 lbs.;
Antimony, 6 lbs.

No. 18. BEST BRITANNIA, *for Spoons.* — Tin, 100 lbs.; Hardening,
5 lbs.; Antimony, 10 lbs.

No. 19. BEST BRITANNIA, *for Handles.* — Tin, 140 lbs.; Copper, 2 lbs.; Antimony, 5 lbs.

No. 20. BEST BRITANNIA, *for Lamps, Pillars, and Spouts.* — Tin, 300 lbs.; Copper, 4 lbs.; Antimony, 15 lbs.

No. 21. CASTING — Tin, 100 lbs; Hardening, 5 lbs.; Antimony, 5 lbs.

No. 22. LINING METAL, *for Boxes of Railroad Cars.* — Mix Tin, 24 lbs.; Copper, 4 lbs.; Antimony, 8 lbs.(for a hardening); then add Tin, 72 lbs.

No. 23. FINE SILVER COLORED METAL. — Tin, 100 lbs.; Antimony, 8 lbs.; Copper, 4 lbs.; Bismuth, 1 lb.

No. 24. GERMAN SILVER, *First Quality for Casting.* — Copper, 50 lbs.; Zinc, 25 lbs.; Nickel, 25 lbs.

No. 25. GERMAN SILVER, *Second Quality for Casting.* — Copper, 50 lbs.; Zinc, 20 lbs.; Nickel, (best pulverized,) 10 lbs.

No. 26. GERMAN SILVER, *for Rolling.* — Copper, 60 lbs.; Zinc, 20 lbs.; Nickel, 25 lbs.

No. 27. GERMAN SILVER, *for Bells and other Castings.* — Copper, 60 lbs.; Zinc, 20 lbs.; Nickel, 20 lbs.; Lead, 3 lbs.; Iron, (that of tin plate being best,) 2 lbs.

No. 28 IMITATION OF SILVER. — Tin, 3 ozs.; Copper, 4 lbs.

No. 29. PINCHBECK. — Copper, 5 lbs.; Zinc, 1 lb.

No. 30. TOMBAC. — Copper, 16 lbs.; Tin, 1 lb.; Zinc, 1 lb.

No. 31. RED TOMBAC. — Copper, 10 lbs.; Zinc, 1 lb.

No 32. HARD WHITE METAL. — Sheet Brass, 32 ozs.; Lead, 2 ozs.; Tin, 2 ozs.; Zinc, 1 oz.

No. 33. METAL FOR TAKING IMPRESSIONS. — Lead, 3 lbs.; Tin, 2 lbs.; Bismuth, 5 lbs.

No. 34. SPANISH TUTANIA. — Iron or Steel, 8 ozs.; Antimony, 16 ozs; Nitre, 3 ozs.
Melt and harden 8 ozs. Tin with 1 oz. of the above compound.

No. 35. ANOTHER TUTANIA. — Antimony, 4 ozs.; Arsenic, 1 oz.; Tin, 2 lbs.

No. 36. GUN METAL. — Bristol Brass, 112 lbs.; Zinc, 14 lbs.; Tin, 7 lbs.

No. 37. RIVET METAL. — Copper, 32 ozs.; Tin, 2 ozs.; Zinc, 1 oz.

No. 38. RIVET METAL, *for Hose.* — Tin, 64 lbs.; Copper, 1 lb.

No. 39. FUSIBLE ALLOY, *(which melts in boiling water.)* — Bismuth, 8 ozs.; Tin, 3 ozs.; Lead, 5 ozs.

No. 40. FUSIBLE ALLOY, *for Silvering Glass.* — Tin, 6 ozs.; Lead, 10 ozs.; Bismuth, 21 ozs.; Mercury, a small quantity.

No. 41. SOLDER, *for Gold.* — Gold, 6 pwts.; Silver, 1 pwt.; Copper, 2 pwts.

No. 42. SOLDER, *for Silver.* — *(For the use of Jewellers.)* — Fine Silver, 19 pwts ; Copper, 1 pwt.; Sheet Brass, 10 pwts.

No. 43. WHITE SOLDER, *for Silver.* — Silver, 1 oz.; Tin, 1 oz.

No. 44. WHITE SOLDER, *for raised Britannia Ware.* — Tin, 100 lbs.; Copper, 3 ozs.; to make it free, add Lead, 3 ozs.

No. 45. BEST SOFT SOLDER, *for Cast Britannia Ware.* — Tin, 8 lbs.; Lead, 5 lbs.

No. 46. YELLOW SOLDER, *for Brass or Copper.* — Copper, 1 lb.; Zinc, 1 lb.

No. 47. YELLOW SOLDER, *for Brass or Copper.* — *(Stronger than the last.)* — Copper, 32 lbs.; Zinc, 29 lbs.; Tin, 1 lb.

No. 48. SOLDER, *for Copper.* — Copper, 10 lbs.; Zinc, 9 lbs.

No. 49. BLACK SOLDER. — Copper, 2 lbs.; Zinc, 3 lbs; Tin, 2 ozs.

No. 50. BLACK SOLDER. — Sheet Brass, 20 lbs.; Tin, 6 lbs.; Zinc, 1 lb.

No. 51. SOFT SOLDER. — Tin, 15 lbs.; Lead, 15 lbs.

No. 52. SILVER SOLDER, *for Plated Metal.* — Fine Silver, 1 oz.; Brass, 10 pwts.

No. 53. YELLOW DIPPING METAL. — Copper, 32 lbs.; Zinc, 2 lbs.; Soft Solder, 2½ ozs.

No. 54. QUICK BRIGHT DIPPING ACID, *for Brass which has been ormoloud.* — Sulphuric Acid, 1 gall.; Nitric Acid, 1 gall.

No. 55. DIPPING ACID. — Sulphuric Acid, 12 lbs.; Nitric Acid, 1 pint; Nitre, 4 lbs.; Soot, 2 handfuls; Brimstone, 2 ozs.
Pulverize the Brimstone and soak it in water an hour. Add the Nitric Acid last.

No. 56. GOOD DIPPING ACID, *for Cast Brass.* — Sulphuric Acid, 1 qt., Nitre, 1 qt.; Water, 1 qt.
A little Muriatic Acid may be added or omitted.

No. 57. DIPPING ACID. — Sulphuric Acid, 4 galls.; Nitric Acid, 2 galls.; Saturated solution of Sulphate of Iron (Copperas), 1 pint; Solution of Sulphate of Copper, 1 qt.

No. 58. ORMOLU DIPPING ACID, *for Sheet Brass.* — Sulphuric Acid, 2 galls ; Nitric Acid, 1 pt.; Muriatic Acid, 1 pt.; Water, 1 pt.; Nitre, 12 lbs.
Put in the Muriatic Acid last, a little at a time and stir the mixture with a stick.

No. 59. ORMOLU DIPPING ACID, *for Sheet or Cast Brass.* — Sulphuric Acid, 1 gall ; Sal Ammoniac, 1 oz.; Sulphur, (in flour,) 1 oz.; Blue Vitriol, 1 oz.; Saturated Solution of Zinc in Nitric Acid, mixed with an equal quantity of Sulphuric Acid, 1 gall.

No. 60. To PREPARE BRASS WORK FOR ORMOLU DIPPING. — If the work is oily, boil it in lye; and if it is finished work, filed or turned, dip it in old acid, and it is then ready to be ormoloed ; but if it is unfinished, and free from oil, pickle it in strong sulphuric acid, dip in pure nitric acid, and then in the old acid, after which it will be ready for ormoloing.

No. 61. To REPAIR OLD NITRIC ACID ORMOLU DIPS. — If the work after dipping appears coarse and spotted, add vitriol till it answers the purpose. If the work after dipping appears too smooth, add muriatic acid and nitre till it gives the right appearance.

The other ormolu dips should be repaired according to the receipts, putting in the proper ingredients to strengthen them. They should not be allowed to settle, but should be stirred often while using.

No. 62. TINNING ACID, *for Brass or Zinc.* — Muriatic Acid, 1 qt., Zinc, 6 ozs. To a solution of this add, Water, 1 qt.; Sal Ammoniac, 2 ozs.

No. 63. VINEGAR BRONZE, *for Brass.* — Vinegar, 10 galls.; Blue Vitriol, 3 lbs.; Muriatic Acid. 3 lbs ; Corrosive Sublimate, 4 grs.; Sal Ammonia, 2 lbs.; Alum, 8 ozs.

No. 64. DIRECTIONS FOR MAKING LACQUER. — Mix the ingredients and let the vessel containing them stand in the sun, or in a place slightly warmed three or four days, shaking it frequently till the gum is dissolved, after which let it settle from twenty-four to forty-eight hours, when the clear liquor may be poured off for use. Pulverized glass is sometimes used in making Lacquer, to carry down the impurities.

No. 65. LACQUER, *for Dipped Brass.* — Alcohol. proof specific gravity not less than 95-100ths, 2 galls.; Seed Lac, 1 lb.; Gum Copal, 1 oz.; English Saffron, 1 oz.; Annotto, 1 oz.

No. 66. LACQUER, *for Bronzed Brass.*—To one pint of the above Lacquer. add. Gamboge, 1 oz.; and after mixing it add an equal quantity of the first Lacquer.

No. 67. DEEP GOLD COLORED LACQUER.—Best Alcohol, 40 ozs.; Spanish Annotto, 8 grs.; Turmeric, 2 drs.; Shell Lac, ½ oz.; Red Sanders, 12 grs.; when dissolved add Spirits of Turpentine, 30 drops.

No. 68. GOLD COLORED LACQUER, *for Brass not Dipped.*—Alcohol, 4 galls.; Turmeric, 3 lbs.; Gamboge, 3 ozs.; Gum Sanderach. 7 lbs.; Shell Lac, 1½ lb.; Turpentine Varnish, 1 pint.

No. 69. GOLD COLORED LACQUER, *for Dipped Brass.*—Alcohol, 36 ozs.; Seed Lac, 6 ozs.; Amber, 2 ozs.; Gum Gutta, 2 ozs.; Red Sandal Wood, 24 grs.; Dragon's Blood, 60 grs.; Oriental Saffron, 36 grs.; Pulverized Glass, 4 ozs.

No. 70. GOOD LACQUER, *for Brass.*—Seed Lac, 6 ozs.; Amber or Copal, 2 ozs.; Best Alcohol, 4 galls.; Pulverized Glass, 4 ozs.; Dragon's Blood, 40 grs.; Extract of Red Sandal Wood obtained by water, 30 grs.

No. 71. LACQUER, *for Dipped Brass.*—Alcohol, 12 galls.; Seed Lac, 9 lbs.; Turmeric, 1 lb. to a gallon of the above mixture; Spanish Saffron, 4 ozs.
The Saffron is to be added for Bronze work.

No. 72. GOOD LACQUER.—Alcohol, 8 ozs.; Gamboge, 1 oz.; Shell Lac, 3 ozs.; Annotto, 1 oz.; solution of 3 ozs. of Seed Lac in 1 pint of Alcohol; when dissolved add ½ oz. Venice Turpentine, ¼ oz. Dragon's Blood, will make it dark; keep it in a warm place four or five days.

No. 73. PALE LACQUER, *for Tin Plate.*—Best Alcohol, 8 ozs.; Turmeric, 4 drs.; Hay Saffron, 2 scs.; Dragon Blood, 4 scs.; Red Sanders. 1 sc.; Shell Lac, 1 oz.; Gum Sanderach, 2 drs.; Gum Mastic, 2 drs.; Canada Balsam, 2 drs.; when dissolved add Spirits of Turpentine, 80 drops.

No. 74. RED LACQUER, *for Brass.*—Alcohol, 8 galls.; Dragon's Blood, 4 lbs.; Spanish Annotto, 12 lbs., Gum Sanderach, 13 lbs.; Turpentine, 1 gall.

No. 75. PALE LACQUER, *for Brass.*—Alcohol, 2 galls.; Cape Aloes cut small, 3 ozs.; Pale Shell Lac, 1 lb.; Gamboge, 1 oz.

No. 76. BEST LACQUER, *for Brass.*—Alcohol, 4 galls.; Shell Lac, 2 lbs.; Amber Gum, 1 lb.; Copal, 20 ozs.; Seed Lac, 3 lbs.; Saffron, to color; Pulverized Glass, 8 ozs.

No 77. COLOR FOR LACQUER.—Alcohol, 1 qt.; Annotto, 4 ozs.

No. 78. LACQUER, *for Pilosophical Instruments.*—Alcohol, 80 ozs.; Gum Gutta, 3 ozs.; Gum Sandarac, 8 ozs.; Gum Elemi, 8 ozs.; Dragon's Blood, 4 ozs; Seed Lac, 4 ozs.; Terra Merita, 3 ozs.; Saffron, 8 grs.; Pulverized Glass, 12 ozs.

No. 79. BROWN BRONZE DIP.—Iron Scales, 1 lb.; Arsenic, 1 oz. Muriatic Acid, 1 lb.; Zinc, (solid,) 1 oz.
Let the Zinc be kept in only while it is in use.

No. 80. GREEN BRONZE DIP.—Wine Vinegar, 2 qts.; Verditer Green, 2 ozs.; Sal Ammoniac, 1 oz ; Salt, 2 ozs.; Alum, ½ oz.; French Berries, 8 ozs.; boil the ingredients together.

No. 81. AQUAFORTIS BRONZE DIP.—Nitric Acid, 8 ozs.; Muriatic Acid, 1 qt.; Sal Ammoniac, 2 ozs.; Alum, 1 oz.; Salt, 2 ozs.; Water, 2 galls.
Add the Salt after boiling the other ingredients, and use it hot.

No. 82. OLIVE BRONZE DIP, *for Brass.* — Nitric Acid, 3 ozs ; Muriatic Acid, 2 ozs.; add Titanium or Palladium ; when the metal is dissolved add 2 galls. pure soft water to each pint of the solution.   •

No. 83. BROWN BRONZE PAINT, *for Copper Vessels.* — Tincture of Steel, 4 ozs. ; Spirits of Nitre, 4 ozs. ; Essence of Dendi, 4 ozs. ; Blue Vitriol, 1 oz.; Water, ½ pint.
Mix in a bottle. Apply it with a fine brush, the vessel being full of boiling water Varnish after the application of the bronze.

No. 84. BRONZE, *for all kinds of Metal.* — Muriate of Ammonia (Sal Ammoniac), 4 drs.; Oxalic Acid, 1 dr.; Vinegar, 1 pint.
Dissolve the Oxalic Acid first. Let the work be clean. Put on the bronze with a brush, repeating the operation as many times as may be necessary.

No. 85 BRONZE PAINT, *for Iron or Brass.* — Chrome Green, 2 lbs.; Ivory Black, 1 oz.; Chrome Yellow, 1 oz.; Good Japan, 1 gill; grind all together and mix with Linseed Oil.

No. 86. TO BRONZE GUN BARRELS.—Dilute Nitric Acid with Water and rub the gun barrels with it ; lay them by for a few days, then rub them with Oil and polish them with bees-wax.

No. 87. FOR TINNING BRASS. — Water, 2 pails full; Cream of Tartar, 1-2 lbs.; Salt, 1-2 pint.
Shaved or Grained Tin. — Boil the work in the mixture, keeping it in motion during the time of boiling.

No. 88. SILVERING BY HEAT. — Dissolve 1 oz. of Silver in Nitric Acid ; add a small quantity of Salt ; then wash it and add Sal Ammoniac, or 6 ozs. of Salt and White Vitriol ; also ¼ oz. of Corrosive Sublimate, rub them together till they form a paste, rub the piece which is to be Silvered with the paste, heat it till the Silver runs, after which dip it in a weak vitriol pickle to clean it.

No. 89. MIXTURE FOR SILVERING. — Dissolve 2 ozs. of Silver with 3 grains of Corrosive Sublimate ; add Tartaric Acid, 4 lbs.; Salt, 8 qts.

No. 90. SEPARATE SILVER FROM COPPER. — Mix Sulphuric Acid, 1 part ; Nitric Acid, 1 part ; Water, 1 part; boil the metal in the mixture till it is dissolved, and throw in a little Salt to cause the Silver to subside.

No. 91. SOLVENT FOR GOLD. — Mix equal quantities of Nitric and Muriatic Acids.

No. 92. VARNISH, *for Smooth Moulding Patterns.* — Alcohol, 1 gall.; Shell Lac, 1 lb.; Lamp or Ivory Black, sufficient to color it.

No. 93. FINE BLACK VARNISH, *for Coaches.* — Melt in an Iron pot, Amber, 32 ozs.; Resin, 6 ozs.; Asphaltum, 6 ozs.; Drying Linseed Oil, 1 pt.; when partly cooled add Oil of Turpentine, wormed, 1 pt.

No. 94. CHINESE WHITE COPPER. — Copper, 40.4 ; Nickel. 31.6 ; Zinc, 25.4 ; and Iron, 2.6 parts.

No. 95. MANHEIM GOLD. — Copper, 3 ; Zinc, 1 part ; and a small quantity of Tin.

No. 96. ALLOY OF THE STANDARD MEASURES USED BY THE BRITISH GOVERNMENT. — Copper, 576 ; Tin, 59 ; and Brass, 48 parts.

No. 97. BATH METAL. — Brass, 32 ; and Zinc, 9 parts.

No. 98. SPECULUM METAL.— Copper, 6 ; Tin, 2 ; and Arsenic, 1 part Or, Copper, 7; Zinc, 3 ; and Tin, 4 parts.

No. 99. HARD SOLDER. — Copper, 2 ; Zinc, 1 part.

No. 100. BLANCHED COPPER. — Copper, 8 ; and Arsenic, ½ part.

No. 101. BRITANNIA METAL. — Brass, 4 ; Tin, 4 parts ; when fused, add Bismuth, 4 ; and Antimony, 4 parts.
This composition is added at discretion to melted Tin.

No. 102. Plumber's Solder. — Lead, 2; Tin, 1 part.

No. 103. Tinman's Solder. — Lead, 1; Tin, 1 part.

No. 104. Pewterer's Solder. — Tin, 2; Lead, 1 part.

No. 105. Common Pewter. — Tin, 4; Lead, 1 part.

No. 106. Best Pewter. — Tin, 100; Antimony, 17 parts.

No. 107. A Metal that Expands in Cooling. — Lead, 9; Antimony, 2; Bismuth, 1 part.

This Metal is very useful in filling small defects in Iron castings, &c.

No. 108. Queen's Metal. — Tin, 9; Antimony, 1; Bismuth, 1; Lead, 1 part.

No. 109. Mock Platinum. — Brass, 8; Zinc, 5 parts.

No. 110. Silver Coin of the United States. — Pure Silver, 9; Alloy, 1 part; the alloy of silver is fine copper.

No. 111. Gold Coin of the United States. — Pure Gold, 9; Alloy, 1 part; the alloy of gold is $\frac{1}{4}$ silver and $\frac{3}{4}$ copper, (not to exceed $\frac{1}{2}$ silver).

No. 112. Silver Coin of Great Britain. — Pure Silver, 11.1; Copper, 9.9 parts.

No. 113. Gold Coin of Great Britain. — Pure Gold, 11; Copper, 1 part.

Previous to 1826, Silver formed part of the alloy of Gold coin; hence the different color of English Gold money.

No. 114. Ring Gold. — Pure Copper, $6\frac{1}{2}$ pwts.; Fine Silver, $3\frac{3}{4}$ pwts.; Pure Gold, 1 oz. and 5 pwts.

No. 115. Mock Gold. — Fuse together Copper, 16; Platinum, 7; Zinc, 1 part.

When Steel is alloyed with 1-500 part of Platinum, or with 1-500 part of Silver, it is rendered much harder, more malleable, and better adapted for every kind of cutting instrument.

Note. — In making alloys, care must be taken to have the more infusible metals melted first, and afterwards add the others.

No. 116. Composition Used in Welding Cast Steel. — Borax, 10; Sal Ammoniac, 1 part; grind or pound them roughly together; then fuse them in a metal pot over a clear fire, taking care to continue the heat until all spume has disappeared from the surface. When the liquid appears clear, the composition is ready to be poured out to cool and concrete; afterwards being ground to a fine powder, it is ready for use.

To use this composition, the Steel to be welded is raised to a heat which may be expressed by "bright yellow;" it is then dipped among the welding powder, and again placed in the fire until it attains the same degree of heat as before, it is then ready to be placed under the hammer.

No. 117. Cast Iron Cement. — Clean borings, or turnings, of Cast Iron, 16; Sal Ammoniac, 2; Flour of Sulphur, 1 part; mix them well together in a mortar and keep them dry. When required for use, take of the mixture, 1; clean borings, 20 parts; mix thoroughly, and add a sufficient quantity of water.

A little grindstone dust added improves the cement.

No. 118. Booth's Patent Grease, for Railway Axles. — Water, 1 gall.; Clean Tallow, 3 lbs.; Palm Oil, 6 lbs.; Common Soda, $\frac{1}{4}$ lb. Or, Tallow, 8 lbs.; Palm Oil, 10.

The mixture to be heated to about 210° F., and well stirred till it cools down to about 70°, when it is ready for use.

No. 119. Cement, for Steam-pipe Joints, &c., with Faced Flanges. — White Lead, mixed, 2; Red Lead, dry, 1 part; grind or otherwise mix them to a consistence of thin putty, apply interposed layers with one or two thicknesses of canvas or gauze with as the necessity of the case may be.

No. 120. Soft Cement, *for Steam-boilers, Steam-pipes, &c.*— Red or White Lead, in oil, 4; Iron borings, 2 to 3 parts.

No. 121. Hard Cement.— Iron Borings and Salt Water, and a small quantity of Sal Ammoniac with Fresh Water.

No. 122. Staining Wood and Ivory.— *Yellow.*— Dilute Nitric Acid will produce it on wood.

Red.— An infusion of Brazil Wood in stale urine, in the proportion of a pound to a gallon for wood ; to be laid on when boiling hot, and should be laid over with alum water before it dries. Or, a solution of Dragon's Blood in spirits of wine, may be used.

Black.—Strong solution of Nitric Acid, for wood or ivory.

Mahogany. — Brazil, Madder, and Logwood, dissolved in water and put on hot.

Blue. — Ivory may be stained thus : Soak it in a solution of Verdigris in Nitric Acid, which will turn it *green ;* then dip it into a solution of Pearlash boiling hot.

Purple. — Soak ivory in a solution of Sal Ammoniac into four times its weight of Nitrous Acid.

TABLE OF ALLOYS.

Alloys having a density greater than the Mean of their Constituents.		Alloys having a density less than the Mean of their Constituents.	
Gold and zinc.	Silver & antimony.	Gold and silver.	Iron and bismuth.
Gold and tin.	Copper and zinc.	Gold and iron.	Iron and antimony.
Gold and bismuth.	Copper and tin.[um	Gold and lead.	Iron and lead.
Gold and antimony.	Copper and palladi-	Gold and copper.	Tin and lead.
Gold and cobalt.	Copper & bismuth.	Gold and iridium.	Tin and palladium.
Silver and zinc.	Lead and antimony	Gold and nickel.	Tin and antimony.
Silver and lead.	Platinum & molyb-	Silver and copper.	Nickel and arsenic.
Silver and tin.	denum. [muth.	Silver and lead.	Zinc and antimony.
Silver and bismuth.	Palladium and bis-		

ALLOYS OF COPPER AND ZINC, AND OF COPPER AND TIN.

Composition by Weight per cent.	Specific Gravity.	Colour.	Ultimate Cohesive Strength of an In. square Bar. in Tons.	Characteristic Properties, &c.
Copper	8667	Tile red.	24.6	Malleable.
100.00 Zinc	6895	Bluish grey.	15.2	Brittle.
83.02+16.98	8415	Yellowish red.	13.7	Bath metal.
79.65+20.35	8413	do. do.	11.7	Dutch brass.
71.58+25.42	8397	Pale yellow.	13.1	Rolled sheet brass.
66.18+33.82	8299	Full yellow.	12.5	British brass.
49.47+50.53	8230	do. do.	9.2	German brass.
32.85+67.15	8283	Deep yellow.	19.3	Watchmakers' brass.
30 30+69.70	7836	Silver white.	2.2	Very brittle.
21.50+75.50	7449	Ash grey.	3.1	Brittle.
19.65+80.35	7371	do.	1.9	White button metal.
Tin	7291	White.	2.7	
84.29+15.71	8561	Reddish yellow.	16.1	Gun metal.
81.10+18.90	8459	Yellowish red.	17.7	Gun metal and bronze.
78.97+21.03	8728	do. do.	13.6	Hard, mill brasses.
34.92+65.08	8065	White.	1.4	Small bells.
15.17+84.83	7447	Very white.	3.1	Speculum metal.
11.82+89.18	7472	do. do.	3.1	Files, tough.

Note.—No simple binary alloy of copper and zinc, or of copper and tin, works as pleasantly in turning, planing, or filing, as if combined with a small proportion of a third fusible metal ; generally lead is added to copper and zinc, and zinc to copper and tin.

9

To POLISH BRASS.— When the Brass is made smooth by turning or filing with a very fine file, it may be rubbed with a smooth fine grained stone, or with charcoal and water.   When it is made quite smooth and free from scratches it may be polished with rotten stone and oil, alcohol or spirits of turpentine.

To CLEAN BRASS.— If there is any oily substance on the Brass boil it in a solution of potash, or strong lye.   Mix equal quantities of Nitric and Sulphuric Acids in a stone or earthern vessel, let it stand a few hours, stirring it occasionally with a stick, then dip the Brass in the solution, but take it out immediately and rinse it in soft water, and wipe it in saw dust till it is dry.

GLUE.— Powdered Chalk added to common Glue strengthens it.   A Glue which will resist the action of water is made by boiling 1 pound of Glue in 2 quarts of skimmed Milk.

ALLOYS FOR BRONZE.

Professor Hoffman, of the Prussian artillery, has made experiments with the view of obtaining a good statuary bronze, and recommends the alloys ranging between the two following admixtures :—

1st.   To produce the reddest bronze.
    88.75 COPPER ZINC (7 atoms copper, 1 atom zinc).
    11.25 COPPER TIN (3 atoms copper, 1 atom tin).

    100·00

2nd.   To produce a cheap bronze, with a bright yellow color, almost golden.
    93.5 COPPER ZINC (2 atoms copper, 1 atom zinc).
    6.5 COPPER TIN (3 atoms copper, 1 atom tin).

    100.0

VALUABLE ALLOYS.

The "Paris Scientific Review" has published, for the benefit of the industrious workers in metals, the best receipts for composing all the various factitious metals used in the arts ; the following are a few :—

STATUARY BRONZE.—Darcet has discovered that this is composed of copper, 91.4; zinc, 5.5 ; lead, 1.7; tin, 1.4.

BRONZE FOR CANNON OF LARGE CALIBRE.—Copper, 90 ; tin, 7.

PINCHBECK.—Copper, 5 ; zinc, 1.

BRONZE FOR CANNON OF SMALL CALIBRE.—Copper, 93 ; tin, 7.

BRONZE FOR MEDALS.—Copper, 100 ; tin, 8.

ALLOY FOR CYMBALS.—Copper, 80 ; tin, 20.

METAL FOR THE MIRRORS OF REFLECTING TELESCOPES.—Copper, 100 ; tin, 50.

WHITE ARGENTAN.—Copper, 8 ; nickel, 3 ; zinc, 35 ; this beautiful composition is in imitation of silver.

CHINESE SILVER.—M. Mairer discovered the following proportions :— Silver, 2.5; copper, 65.24; zinc, 19.52; nickel, 13; cobalt of iron, 0.12.

TUTENAG.—Copper, 8 ; nickel, 3 ; zinc, 5.

PRINTING CHARACTERS.—Lead, 4 ; antimony, 1.   For stereotype plates—Lead, 9 ; antimony, 2 ; bismuth, 2.

MECHANICAL DRAWING

AND

INSTRUMENTS USED IN DRAWING.

·

INSTRUMENTS USED IN DRAWING.

To facilitate the construction of geometrical figures, we add a short description of a few useful instruments which do not belong to the common pocket-case.

Let there be a flat ruler, AB, from one to two feet in length, for which the common Gunter's scale may be substituted; and, secondly, a triangular piece of wood, a, b, c, flat, and about the same thickness as the ruler: the sides, ab and bc, of which are equal to one another, and form a right angle at b. For the convenience of sliding, there is usually a hole in the middle of the triangle, as may be seen in the figure.

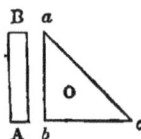

By means of these simple instruments many very useful geometrical problems may be performed. Thus, to draw a line through a given point parallel to a given line. Lay the triangle on the paper so that one of its sides will coincide with the given line to which the parallel is to be drawn; then, keeping the triangle steady, lay the ruler on the paper, with its edge applied to either of the other sides of the triangle; then, keeping the ruler firm, move the triangle along its edge, up or down, to the given point; the side of the triangle which was placed on the given line will always keep parallel to itself, and hence a parallel may be drawn through the given point.

To erect a perpendicular on a given line, and from any given point in that line, we have only to apply the ruler to the given line, and place the triangle so, that its right angle shall touch the given point in the line, and one of the sides about the right angle, placed to the edge of the ruler—the other side will give the perpendicular required.

If the given point be either above or below the line, the process is equally easy. Place one of the sides of the triangle about the right angle on the given line, and the ruler on the side opposite the right angle, then slide the triangle on the edge of the ruler till the given point from which the perpendicular is to be drawn is on the other side, then this side will give the perpendicular.

Other problems may be performed with these instruments, the method of doing which it will be easy for the reader to contrive for himself.

When arcs of circles of great diameter are to be drawn, the use of a compass may be substituted by a very simple contrivance. Draw the chord of the arc to be described, and place a pin at each extremity, A and B, then place two rulers jointed at C, and forming an angle, ACB equal to the supplement of half the given number of degrees; that is to say, the number of degrees which the arc whose chord given is to contain, is to be halved, and this half being subtracted from 180 degrees, will give the degrees which form the angle at which the rulers are placed, that is, the angle ACB. This being done, the

9*

edges of the rulers are moved along against the pins, and a pencil at C will describe the arc required.

Large circles may be described by a contrivance equally simple. On an axle, a foot or a foot and a half long, there are placed two wheels, M and F, of which one is fixed to the axle, namely F, and the other is capable of being shifted to different parts of the axle, and, by means of a thumb-screw, made capable of being fixed at any point on the axle. These wheels are of different diameters, say of 3 and 6 inches, the fixed wheel F being the largest. This instrument being moved on the paper, the circles M and F will roll, and describe circles of different radii: the axle will always point to the centre of these circles, and there will be this proportion:

As the diameter of the large wheel is to the difference of the diameters of the two wheels, so is the radius of the circle to be described by the large wheel to the distance of the two wheels on the axle.

If the diameters of the wheels are as above stated, and it is required to describe a circle of 3 feet radius, then from the above proportion we have $6 : 6 — 3 :: 3$ feet or 36 inches : 18 inches $=$ the distance of the two wheels, to describe a circle 6 feet in diameter.

It may be observed, that it will be best to make the difference of the wheels greater if large circles are to be described, as then a shorter instrument will serve the purpose.

We will conclude these instructions, by making a few remarks on the Diagonal Scale and Sector, the great use of the latter of which, especially, is seldom explained to the young mechanic.

The diagonal scale to be found on the plain scale in common pocket-cases of instruments, is a contrivance for measuring very small divisions of lines; as, for instance, hundredth parts of an inch.

Suppose the accompanying cut to represent an enlarged view of two divisions of the diagonal scale, and the bottom and top lines to be divided into two parts, each representing the tenth part of an inch. Now, the perpendicular lines BC, AD, are each divided into ten equal parts, which are joined by the crossing lines, 1, 2, 3, 4, &c., and the diagonals BF, DE, are drawn as in the figure. Now, as the division FC is the tenth part of an inch, and as the line FB continually approaches nearer and nearer to BC, till it meets it in B, it will follow, that the part of the line 1 cut off by this diagonal will be a tenth part of FC, because B1 is only one-tenth part of BC; so, likewise, 2 will represent two-tenth parts, 3 three-tenth parts, and so on to 9, which is nine-tenth parts, and 10, ten-tenth parts, or the whole tenth of an inch; so that, by means of this diagonal, we arrive at divisions equal to tenth parts of tenth parts of an inch, or hundredths of an inch. With this consideration, an examination of the scale itself will easily show the whole matter. It may be observed,

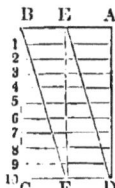

that if half an inch and the quarter of an inch be divided, in the same manner, into tenths and tenths of tenths, we may get thus two-hundredth and
four-hundredth parts of an inch.

THE SECTOR.

This very useful instrument consists of two equal rulers each six inches
long, joined together by a brass folding joint. These rulers are generally
made of boxwood or ivory; and on the face of the instrument, several lines
or scales are engraven. Some of these lines or scales proceed from the
centre of the joint, and are called *sectorial lines*, to distinguish them from
others which are drawn parallel to the edge of the instrument, similar to
those on the common Gunter's scale.

The sectorial lines are drawn twice on the same face of the instrument;
that is to say, each line is drawn on both legs. Those on each face are,

A scale of equal parts, marked L,
A line of chords, marked   C,
A line of secants, marked   S,
A line of polygons, marked   P, or Pol.

These sectorial lines are marked on one face of the instrument; and on the
other there are the following;

A line of sines, marked   S,
A line of tangents, marked   T,
A line of tangents to a less radius, marked t.

This last line is intended to supply the defect of the former, and extends
from about 45 to 75 degrees.

The lines of chords, sines, tangents, and secants, but not the line of polygons, are numbered from the centre, and are so disposed as to form equal
angles at the centre; and it follows from this, that at whatever distance the
sector is opened, the angles which the lines form, will always be respectively
equal. The distance, therefore, between 10 and 10, on the two lines marked
L, will be equal to the distance of 60 and 60 on the two lines of chords, and
also to 90 and 90 on the two lines of sines, &c. at any particular opening of
the sector.

Any extent measured with a pair of compasses, from the centre of the
joint to any division on the sectorial lines, is called a *lateral distance;* and
any extent taken from a point in a line on the one leg, to the like point on
the similar line on the other leg, is called a *transverse* or *parallel distance.*

With these remarks, we shall now proceed to explain the use of the sector, in so far as it is likely to be serviceable to mechanics.

USE OF THE LINE OF LINES.

This line, as was before observed, is marked L, and its uses are,

To Divide a line into any number of equal parts: Take the length of the
line by the compasses, and placing one of the points on that number in the

line of lines which denotes the number of parts into which the given line is to be divided, open the sector till the other point of the compasses touches the same division on the line of lines marked on the other leg; then, the sector being kept at the same width, the distance from 1 on the line L on the one leg, to 1 on the line L on the other, will give the length of one of . the equal divisions of the given line to be divided. Thus, to divide a given line into seven equal parts :—take the length of the given line with the compasses, and setting one point on 7, on the line L of one of the legs, move the other leg out until the other point of the compasses touch 7 on the line L of that leg; this may be called the transverse distance of 7 on the line of lines. Now, keeping the sector at the same opening, the transverse distance of 1 will be the length of one of the 7 equal divisions of the given line; the transverse distance of 2 will be two of these divisions, &c.

It will sometimes happen, that the line to be divided will be too long for the largest opening of the sector; and in this case we take the half, or third, or fourth of the line, as the case may be; then the transverse distance of 1 to 1, will be a half, a third, or a fourth of the required equal part.

To divide a given line into any number of parts that shall have a certain relation or proportion to each other : Take the length of the whole line to be divided, and placing one point of the compasses at that division on the line of lines on one leg of the instrument which expresses the sum of all the parts into which the given line is to be divided, and open the sector till the other point of the compasses is on the corresponding division on the line of lines of the other leg. This is evidently making the sum of the parts into which the given line is to be divided a transverse distance ; and when this is done, the proportional parts will be found by taking, with the same opening of the sector, the transverse distances of the parts required.—To divide a given line into three parts, in the proportion of 2, 3, 4: The sum of these is 9 ; make the given line a transverse distance between 9 and 9 on the two lines of lines ; then the transverse distances of the several numbers 2, 3, 4, will give the proportional parts required.

To find a fourth proportional to three given lines : take the lateral distance of the second, and make it the transverse distance of the first, then will the transverse distance of the third be the lateral distance of the fourth; then, let there be given 6 : 3 : : 8,—make the lateral distance of 3 the transverse distance of 6; then will the transverse distance of 8 be the lateral distance of 4, the fourth proportional required.

This sector will be found highly serviceable in drawing plans. For instance, if it is wished to reduce the drawing of a steam engine from a scale of $1\frac{1}{2}$ inches to the foot, to another of five-eighths to the foot. Now, in $1\frac{1}{2}$ inches there are 12 eighth parts ; so that the drawing will be reduced in the proportion of 12 to 5. Take the lateral distance of 5, and keep the compasses at this opening; then open the sector till the points of the compasses mark the transverse distance of 12; keep now the sector at this opening, and any measure taken on the drawing, to be copied and laid off on the sector as a lateral distance,—the transverse distance taken from that point will give the corresponding measure to be laid down in the new drawing.

If the length of the side of a triangle, of which we have the drawing, is to be reckoned 45; what are the lengths of the other two sides? Take the length of the side given, by the compasses, and open the sector till the measure be the transverse distance of 45 to 45; then the lengths of the other sides being applied transversely, will give their numerical lengths.

USE OF THE LINE OF CHORDS.

By means of the sector, we may dispense with the protractor. Thus, to lay down an angle of any number of degrees :—take the radius of the circle on the compasses, and open the sector till this becomes the transverse distance of 60 on the line of chords; then take the transverse distance of the required number of degrees, keeping the sector at the same opening; and this transverse distance being marked off on an arc of the circle whose radius was taken, will be the required number of degrees.

We will not enter farther on the use of the sectorial lines, as what we have said will, we hope, be found sufficient for the purposes of the practical mechanic.

———

MECHANICAL DRAWING AND PERSPECTIVE.

A FLAT rectangular board is first to be provided, of any convenient size, as from 18 to 30 inches, and from 16 to 24 inches broad. It may be made of fir, plane tree, or mahogany; its face must be planed smooth and flat, and the sides and ends as nearly as possible at right angles to each other— the bottom of the board and the left side should be made perfectly so; and this corner should be marked, so that the stock of the square may be always applied to the bottom and left hand side of the board. To prevent the board from casting, it is usual to pannel it on the back or on the sides.

A T square must also be provided, which by means of a thumb-screw fixed in the stock, may be made to answer either the purposes of a common square, or bevel,—the one-half of the stock being movable about the screw, and the other fixed at right angles on the blade. The blade ought to be somewhat flexible, and equal in length to the length of the board.

Besides these, there will be required a case of mathematical instruments; in the selection of which it should be observed, that the bow compass is more frequently defective than any of the other instruments. After using any of the ink feet, they should be dried; and if they do not draw properly, they ought to be sharpened and brought to an equal length in the blade, by grinding on a hone.

The colors most useful are, Indian ink, gamboge, Prussian blue, vermilion, and lake. With these, all colors necessary for drawing machinery or buildings may be made; so that, instead of purchasing a box of colors, we would advise that those for whom this book is intended should procure these cakes separately : the gamboge may be bought from an apothecary—a pennyworth will serve a lifetime. In choosing the rest, they should be rubbed against the teeth, and those which feel smoothest are of the best quality.

Hair pencils will also be necessary, made of camel's hair, and of various sizes. They ought to taper gradually to a point when wet in the mouth, and should, after being pressed against the finger, spring back.

Black-lead pencils will also be necessary. They ought not to be very soft, nor so hard that their traces cannot be easily erased by the Indian rubber. In choosing paper, that which will best suit this kind of drawing is thick, and has a hardish feel, not very smooth on the surface, yet free from knots.

The paper on which the drawing is to be made, must be chosen of a good quality and convenient size. It is then to be wet with a sponge and clean water, on the opposite side from that on which the drawing is to be made. When the paper absorbs the water, which may be seen by the wetted side becoming dim, as its surface is viewed slantwise against the light, it is to be laid on the drawing board with the wetted side next the board. About half an inch must be turned up on a straight edge all round the paper, and then fastened on the board. This is done because the paper when wet is enlarged, and the edges being fixed on the board, act as stretchers when the paper contracts by drying. To prevent the paper from contracting before the paste has been sufficiently fastened by drying, the paper is usually wet on the upper surface, to within half an inch of the paste mark. When the paper is thoroughly dried, it will be found to lie firmly and equally on the board, and is then fit for use.

If the drawing is to be made from a copy, we ought first to consider what scale it is to be drawn to. If it is to be equal in size to, or larger than the copy, a scale should be made accordingly, by which the dimensions of the several parts of the drawing are to be regulated. The diagonal scale, a simple and beautiful contrivance, will be here found of great use for the more minute divisions ; and whenever the drawing is to be made to a scale of 1 inch, $\frac{1}{2}$ inch, $\frac{1}{4}$ inch to the foot, a scale should be drawn of 20 or 30 equal parts ; the last of which should be subdivided into 12, and a diagonal scale formed on the same principles as the common one, but with eight parallels and 12 diagonals, to express inches and eighths of an inch. For making such scales to any proportion, the line L on the sector will be found very convenient.

Great care should be taken in the penciling, that an accurate outline be drawn, for on this much of the value of the picture will depend. The pencil marks should be distinct, yet not heavy, and the use of the rubber avoided as much as possible, as its frequent application ruffles the surface of the paper. The methods already given for constructing geometrical figures will be here found applicable, and the use of the T square, parallel ruler, &c., will suggest themselves whenever they require to be employed.

The drawing thus made of any machine or building is called a plan. Plans are of three kinds—a ground plan, or bird's-eye view, an elevation or front view, and a perspective plan.

When a view is taken of the teeth of a wheel, with the circumference towards the eye, the teeth appear to be nearer as they are removed from the middle point of the circumference opposite the eye, and it may not be out of place here to give the method of representing them on paper :—If AB be the circumference of a wheel as viewed by the eye, and it is required to represent the teeth as they appear on it, only half of the circumference can be seen in this way at one time, consequently we can only represent the half of the teeth. On AB describe a semicircle, which divide into half as many equal parts as the wheel has teeth; then from each of these points of division draw perpendiculars to the wheel AB, then will these perpendiculars mark the relative places of the teeth.

When the outline is completed in pencil, it is next to be carefully gone over with Indian ink, which is to be rubbed down with a little water, on a plate of glass or eathernware—so as to be sufficiently fluid to flow easily out of the pen, and at the same time have a sufficient body of color. While drawing the ink lines, the measurements should be repeated, so as to correct any error that may have occurred during the penciling. The screw in the drawing pen will regulate the breadth of the strokes; which should not be alike heavy; those strokes being the heaviest which bound the dark part of the shades. Should any line be wrong drawn with the ink, it may be taken out by means of a sponge and water, which could not be done if common writing ink were employed.

In preparing for coloring it is to be observed, that a hair pencil is to be fixed at each end of a small piece of wood, made in the form of a common pencil, one of which is to be used with color, and the other with water only. If the color is to be laid on, so as to represent a flat surface, it ought to be spread on equally, and there is here no use for the water brush; but if it is to represent a curved surface, then the color is to be laid on the part intended to be shaded, and softened towards the light by washing with the water brush. In all cases it should be borne in mind, that the color ought to be laid on very thin, otherwise it will be more difficult to manage, and will never make so fine a drawing.

In colors even of the best quality, we sometimes meet with gritty particles, which it is desirable to avoid. Instead of rubbing the color on a plate with a little water, as is usual, it will be better to wet the color, and rub it on the point of the forefinger, letting the dissolved part drop off the finger on to the plate.

In using the Indian ink, it will be found advantageous to mix it with a little blue and a small quantity of lake, which renders it much more easily wrought with, and this is the more desirable as it is the most frequently used of all the other colors in Mechanical Drawing, the shades being all made with this color.

The depth and extent of the shades will depend on various circumstances—on the figure of the object to be shaded, the position of the eye of the observer, and the direction in which the light comes, &c. The position of the eye will vary the proportionate size of any object in a picture when drawn in perspective. Thus, if a perspective view of a steam engine is given, the eye being supposed to be placed opposite the end nearest the nozzles, an inch of the nozzle rod will appear much larger than an inch of the pump rod which feeds the cistern; but if the eye is supposed to be placed opposite the other end of the engine, the reverse will be the case. But in drawing elevations and ground plans of machinery, every part of the machine is drawn to the proper scale—an inch or foot in one part of the machine, being just the same size as an inch or foot in any other part of the machine. So that by measuring the dimensions of any part of the drawing, and then applying the compass to the scale, we determine the real size of the part so measured. Whereas, if the view were given in perspective, we would be obliged to make allowance for the effect of distance, &c.

The light is always supposed to fall on the picture at an angle of forty-five degrees, from which it follows, that the shade of any object, which is intended to rise from the plane of the picture, or appear prominent, will just be equal in length to the prominence of the object.

The shades, therefore, should be as exactly measured as any other part of the drawing, and care should be taken that they all fall in the proper direction, as the light is supposed to come from one point only.

It is frequently of great use for the mechanic to take a hasty copy of a drawing, and many methods have been given for this purpose—by machines, tracing, &c. We give the following as easy, accurate, and convenient.

Mix equal parts of turpentine and drying oil, and with a rag lay it on a sheet of good silk paper, allowing the paper to lie by for two or three days to dry, and when it is so it will be fit for use. To use it, lay it on the drawing to be copied, and the prepared paper being nearly transparent, the lines of the drawing will be seen through it, and may be easily traced with a black-lead pencil. The lines on the oiled paper will be quite distinct when it is laid on white paper. Thus, if the mechanic has little time to spare, he may take a copy and lay it by to be recopied at his leisure.

Care and perseverance are the chief requisites for attaining perfection in this species of drawing. Every mechanic should know something of it, so that he may the better understand how to execute plans that may be submitted to him, or make intelligible to others any invention he himself may make.

PRACTICAL GEOMETRY.

Geometry is the science which investigates and demonstrates the properties of lines on surfaces and solids: hence, PRACTICAL GEOMETRY is the method of applying the rules of the science to practical purposes.

10

DEFINITION OF ARITHMETICAL SIGNS USED IN THE WORK.

= When we wish to state that one quantity or number, is equal to another quantity or number, the sign of *equality* = is employed. Thus 3 added to 2 = 5, or 3 added to 2 is equal to 5.

+ When the sum of two quantities or numbers is to be taken, the sign *plus* + is placed between them. Thus 3 + 2 = 5; that is, the sum of 3 and 2 is 5. This is the sign of Addition.

— When the difference of two numbers or quantities is to be taken, the sign *minus* — is used, and shows that the latter number or quantity is to be taken from the former. Thus 5 — 2 = 3. This is the sign of Subtraction,

× When the product of any two numbers or quantities is to be taken. the sign *into* × is placed between them. Thus 3 × 2 = 6. This is the sign of Multiplication.

÷ When we are to take the quotient of two quantities, the sign *by* ÷ is placed between them, and shows that the former is to be divided by the latter. Thus 6 ÷ 2 = 3. This is the sign of Division. But in some cases in this work, the mode of division has been, to place the dividend above a horizontal line, and the divisor below it, in the form of a vulgar fraction, thus :

$$\frac{\text{Dividend}}{\text{Divisor}} = \text{Quotient.} \qquad \frac{6}{2} = 3.$$

When the square of any number or quantity is to be taken, this is denoted by placing a small figure 2 above it to the right. Thus 6^2 shows that the square of 6 is to be taken, and therefore $6^2 = 6 \times 6 = 36$.

When we wish to show that the square root of any number or quantity is to be taken, this is denoted by placing the *radical sign* $\surd$ before it. Thus $\surd 36$ shows that the square root of 36 ought to be taken, hence $\surd 36 = 6$.

The common marks of proportion are also used, viz. : :: : as 3 : 6 :: 4 : 8, being read 3 is to 6 as 4 is to 8.

The application of these signs to the expression of rules is exceedingly simple. Thus, connected with the circle we have the following rules :

1st. The circumference of a circle will be found by multiplying the diameter by 3·1416.

2d. The diameter of a circle may be found by dividing the circumference by 3·1416.

3d. The area of a circle may be found by multiplying the half of the diameter, by the half of the circumference, or by multiplying together the diameter and circumference, and dividing the product by 4, or by squaring the diameter and multiplying by ·7854.

Now all these rules may be thus expressed :

1st.   diameter × 3·1416 = circumference.

2d. $\dfrac{\text{circumference}}{3 \cdot 1416} = \text{diameter.}$

3d. $\dfrac{\text{diameter}}{2} \times \dfrac{\text{circumference}}{2} = \text{area.}$

or, $\dfrac{\text{diameter} \times \text{circumference}}{4} = \text{area.}$

or, $\text{diameter}^2 \times \cdot 7854 = \text{area.}$

PRACTICAL GEOMETRY.

——— ·

Practical Geometry is an important branch of knowledge to all who are in any way engaged in the art of building. The workman, as well as the designer, requires its aid; and unless he is acquainted with some of the leading principles of the science, he will frequently feel an uncertainty as to the results he may deduce from the problems which are presented to his notice.

———

PROBLEM I.

To inscribe an Equilateral Triangle within a given Circle.

Let A B C be a circle; it is required to draw within it a triangle

FIG. 1.

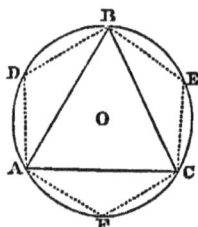

whose sides are equal to one another. Commencing from any point A, mark on the circumference of the circle a series of spaces equal to the radius of the circle, of which there will be six, and draw the arcs A D D B, &c. Then join every alternate point as A B, B C, C A, and the several lines will together form an equilateral triangle.

PROBLEM II.

Within a given Circle to inscribe a Square.

Let A B C D be the given circle, it is required to draw a square

FIG. 2.

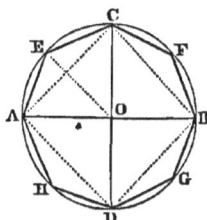

within it.　Draw the diameters A B, C D, at right angles to each other; or, in other words, draw the diameter A B, and form a perpendicular bisecting it.　Then join the points A C, C B, B D, D A, and the figure A B C D is a square formed within a given circle.

PROBLEM III.

Within a given Circle to inscribe a regular Pentagon ; that is, a Polygon of five Sides.

Let A B C D be a circle in which it is required to draw a pentagon.

FIG. 3.

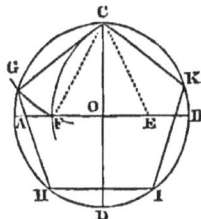

Draw a diameter A D, and perpendicular to it another diameter. Then divide O B into two equal parts in the point E, and join C E ; and with E as a centre, and the radius C E, draw the arc C F, cutting A O in F : and, with C as a centre, and the same radius, describe the arc F G ; the arcs C F, G F intersect each other in the point F, and the arc G F intersects the circumference of the circle in the point G. Join the points C and G, and that line will be a side of the pentagon to be drawn.　Mark off within the circumference the same space, and join the points A H, H I, I K, K C, and the figure that is formed is a pentagon.

PROBLEM IV.

Within a given Circle to describe a regular Hexagon ; that is to say, a Polygon of six equal Sides.

Let A B C be the given circle, and o the centre.   With the radius

FIG. 4.

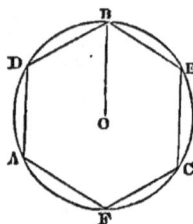

of the circle divide it into parts, of which there will be six, and con-
nect the points A D, D B, &c., and the figure A D B E C F will be a
regular hexagon.

PROBLEM V.

To cut off the Corners of a given Square, so as to form a regular Octagon.

Let A B C D be the given square.   Draw the two diagonal lines

FIG. 5.

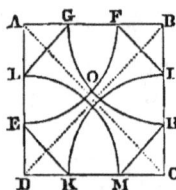

A C, and B D, crossing each other in o.   Then, with the radius A o,
that is, half the diagonal, and with A as a centre, describe the arc
E F, cutting the sides of the square in E and F; then, from B as a
centre, describe the arc G H; and in like manner from C and D describe the arcs I K and L M.   Draw the lines L G, F I, H M, and
K E, and these, with the parts of the given square G F, L H, M K,
and E L, form the octagon required.

10*

PROBLEM VI.

To divide a given Line into any Number of Parts, which Parts shall be in the same Proportion to each other as the Parts of some other given Line, whether those Parts are equal or unequal.

Let A B be the given line which it is required to divide in the same

FIG. 6.

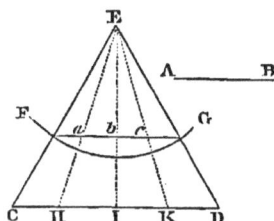

manner and proportion as the line C D, whether the parts are equal or unequal. On the base line C D, form an equilateral triangle in the manner already described in a former problem. Then take the distance A B, and with E as a centre, describe the arc F G, and join the points F and G, and F G shall be equal to A B. Now, if from the points H I K, which are the divisions of the line C, we draw lines to E, as H E, I E, and K E, these lines will cut F G in the points a b c, which will divide the line F G into parts proportionate to the divisions of the line C D.

PROBLEM VII.

On a given Line to draw a Polygon of any Number of Sides, so that that Line shall be one Side of a Polygon; or, in other words, to find the Centre of a Circle which shall circumscribe any Polygon, the Length of the Side of the Polygon being given.

We shall here show, in a tabular form, the length of the radius of a circle, which shall contain the given line, as a side of the required polygon; and here we will suppose the line to be divided into one thousand equal parts, and the radius into a certain number of like parts. The radius of the circle for different figures will be as follows: —

For an inscribed Triangle 577
            Square 701
            Pentagon 850
            Hexagon 1000
            Heptagon 1152
            Octagon 1306½
            Enneagon 1462

Decagon 1618
Endecagon 1775
Dodecagon 1932

By this table, the workman may, with a simple proportion, find the radius of a circle which shall contain a polygon, one side being given : thus, if it be required to draw a pentagon, the side given being fifteen inches, we may say as 1000 is to 15, so is 850, the tabular number for a pentagon, to 12 inches and seventy-five hundredth parts of an inch, or seven-tenths and a half of a tenth of an inch.

We may here give another table for the construction of polygons, one in which the radius of the circumscribing circle is given. If it be required to find the side of the inscribed polygon, the radius being one thousand parts, the sides of the different polygons will be according to the following scale : —

The Triangle 1732
Square 1414
Pentagon 1175
Hexagon 1000
Heptagon 867½
Octagon 765
Enneagon 684
Decagon 618
Endecagon 563½
Dodecagon 517½

Here, as in the case already mentioned, the law of proportion applies, and the statement may be thus made : as one thousand is to the number of inches contained in the radius of the given circle, so is the tabular number for the required polygon to the length of one of its sides in inches. Thus, let it be supposed that we have a circle whose radius in inches is 30, and that we wish to inscribe an octagon within it ; then say as 1000 is to 30 inches, so is 765 to 22 inches and 95-100 parts of an inch, the length of the side of the required octagon.

METHOD OF DRAWING CURVED LINES.

We will now introduce a few remarks upon the method of drawing curved lines, and also give some rules for finding the forms of mouldings when they are to mitre together, that is to say, of raking mouldings, and of bevel work in general. It will also be necessary to make a few remarks upon the form of ribs for domes and groins, a knowledge of which is so necessary to the builder, that without it the workman cannot correctly execute his task. It is hardly necessary to state, that all these mechanical operations are founded upon geometrical principles ; and, unless he is acquainted with these, the workman cannot hope to succeed in his attempt to excel in his art,— one which is necessary for the comfort and convenience of all communities.

Problem VIII.

To draw on Ellipse with the Rule and Compasses, the transverse and conjugate Diameters being given; that is, the Length and Width.

Let A B be the transverse or longest diameter; C D the conjugate

Fig. 7.

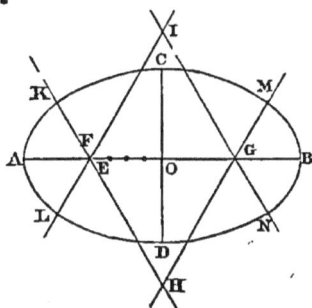

or shortest diameter; and o the point of their intersection, that is, the centre of the ellipse. Take the distance o c or o d; and, taking A as one point, mark that distance A E upon the line A o. Divide o E into three equal parts, and take from A F, a distance E F, equal to one of those parts. Make o G equal to o F. With the radius F G, and F and G as centres, strike arcs which shall intersect each other in the points I and H. Then draw the lines H F K, H G M, and I F L, I G N. With F as a centre, and the radius A F, describe the arc L A K; and, from G as a centre, with the same radius, describe the arc M B N. With the radius H C, and H as a centre, describe the arc K C M; and, from the point I, with the radius I D, describe the arc L D M. The figure A C B D is an ellipse, formed of four arcs of circles.

Problem IX.

To draw an Ellipse by means of two Concentric Circles.

Fig. 8.

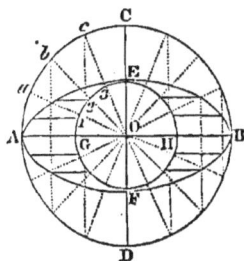

Let A B be the transverse, and E F the conjugate diameter, and o the centre of an ellipse to be drawn. From o with the radius o A, describe the circle A C B D, and from the same centre describe another circle G E H F. Divide the outer circle into any number of equal parts; the greater the number, the more exact will be the ellipse : and they should not be less than twelve. From each of these divisions draw lines to the centre o, as a o, b o, c o. Then, from a, b, c, &c., draw lines perpendicular to A B, and from the corresponding points in the inner circle, that is, from the points marked 1, 2, 3, &c., draw lines parallel to A B. Draw a curve through the points where these lines intersect each other, and it will be an ellipse.

In the diagram to which this demonstration refers, only one quarter of the ellipse is lettered, but the process described in relation to that must be carried round the circles, as is shown in the dotted and other lines.

PROBLEM X.

To describe an Ellipse by Means of a Carpenter's Square, or a piece of notched Lath.

Having drawn two lines to represent the diameters of the ellipse required, fasten the square so that the internal angle or meeting of the blade and stock shall be at the centre of the ellipse. Then take a piece of wood or a lath, and cut it to the length of half the longest diameter, and from one end cut out a piece equal to half the shortest diameter, and there will then be a piece remaining at one end equal to the difference of the half of the two diameters. Place this projecting piece of the lath in such a manner that it may rest against the square, on the edge which corresponds to the two diameters ; then, turning it round horizontally, the two ends of the projection will slide along the two internal edges of the square, and if a pencil be fixed at the other end of the lath, it will describe one quarter of an ellipse. The square must then be moved for the successive quarters of the ellipse, and the whole figure will thus be easily formed.

This method of forming an ellipse is a good substitute for the usual plan, and the figure thus produced is more accurate than that made by passing a pencil round a string moving upon two pins or nails fixed in the foci, for the string is apt to stretch, and the pencil cannot be guided with the accuracy required.

There are many other methods of drawing ellipses, or more properly ovals, but we can only notice two of those in common use.

1. By ordinates, or lines drawn perpendicular to the axis. Having formed the two diameters, divide the axis, or larger diameter, into any number of equal parts, and erect lines perpendicular to the several points. Next draw a semicircle, and divide its diameter into the like number of equal parts; that is, if the larger diameter or axis of the intended ellipse be divided into twenty equal parts, then the

semicircle must be divided into the like number. As the diameter of the semicircle is equal to the shorter diameter of the ellipse, or conjugate axis, perpendiculars may be raised from these divisions of the diameter, or the semicircle, till they meet the circumference; and the different perpendiculars, which are called ordinates, may be erected like perpendiculars, on the axis of ellipse. Joining the several points together, the ellipse is described; and the more accurately the perpendiculars are formed the more exact will be the ellipse.

2. 'By intersecting arches. Take any point in the axis, and with a radius equal to the distance of that point from one extremity of the axis, and with one of the foci as a centre, describe an arc; then with the distance of the assumed point in the axis from the other end of it, and with the other focus as a centre, describe another arc intersecting the former, and the point of intersection will be a point in the ellipse. By assuming any number of points in the axis, any number of points on the curve may be found, and these united will give the ellipse. This process is founded on the property of the ellipse; that if any two lines are drawn from the foci to any point in the curve, the length of these lines added together will be a constant quantity, that is, always the same in the same ellipse.

PROBLEM XI.

To find the Centre and the two Axes of an Ellipse.

Let A B C D be an ellipse, it is required to find its centre. Draw

FIG. 9.

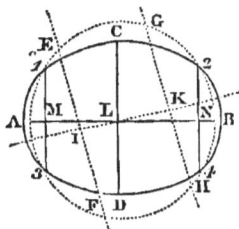

any two lines, as E F and G H, parallel and equal to each other. Bisect these lines as in the points I and K, and bisect I K as in L. From L, as a centre, draw a circle cutting the ellipse in four points, 1, 2, 3, 4. Now L is the centre of the ellipse. But join the points 1, 3, and 2, 4; and bisect these lines as in M and N. Draw the line M N, and produce it to A and B, and it will be the transverse axis. Draw C D through L, and perpendicular to A B, and it will be the conjugate or shorter axis.

Problem XII.

*To draw a flat Arch by the intersection of Lines, having the Open-
ing and Spring or Rise given.*

Let A D B be the opening, and C D its spring or rise. In the mid-

Fig. 10.

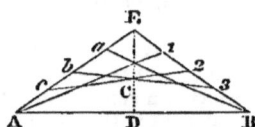

dle of A B, at D, erect a perpendicular D E, equal to twice C D, its
rise ; and from E draw E A and E B, and divide A E and B E into any
number or equal parts, as *a, b, c,* and 1, 2, 3. Join B *a,* 3 *c,* 2 *b,* and
1 A, and it will form the arch required.

The more parts A E and B E are divided into, the greater will be
the accuracy of the curve.

Many curves may be made in the same manner, according to the
position of the lines A E and E B ; and if instead of two lines drawn
from A and B, meeting in E, a perpendicular be erected at the same
points, and two lines be then drawn from the ends of these perpendiculars meeting in an angle, and these lines be divided into any number of equal parts, the points of the adjacent lines may be joined, and
a curve will be formed resembling a gothic arch. The demonstration
already given is therefore very useful to the workman, as he may
vary the form of the curve by altering the position of the lines, either
with respect to the angles which they make with each other, or their
proportional lengths.

Problem XIII.

*To find the Form or Curvature of a raking Moulding that shall
unite correctly with a level one.*

Let A B C D be part of the level moulding, which we will here

Fig. 11.

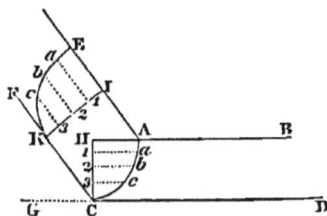

suppose to be an ovolo, or quarter round ; A and C, the points where
the raking moulding takes its rise on the angle ; F C G, the angle the

raking moulding makes with the horizontal one.   Draw c f at the
given angle, and from A draw A E parallel to it ; continue B A to H,
and from c make c H perpendicular to A H.   Divide c H into any
number of equal parts, as 1, 2, 3, and draw lines parallel to H A, as 1
a, 2 *b*, 3 *c ;* and then in any part of the raking moulding, as 1, draw
I K perpendicular to E A, and divide I K into the same number of
equal parts H c is divided into ; and draw 1 *a*, 2 *b*, 3 *c*, parallel to E A.
Then transfer the distances 1 *a*, 2 *b*, 3 *c*, and a curve drawn through
these points will be the form of the curve required for the raking
moulding

We have here shown the method to be employed for an ovolo ; but
it is just the same for any other formed moulding, as a cavetto, semi-
recta, &c.   It may be worthy remark, that, after the moulding is
worked, and the mitre is cut in the mitre-box for the level moulding,
the raking moulding must be cut, either by the means of a wedge
formed to the required angle of the rake, or a box made to correspond
to that angle : and if this be accurately done, the mitre will be true,
and the moulding in all its members correspond to the level moulding.
The plane in which the raking moulding is situated is square to that
of the level one.   This is always the case in a pediment, the mouldings of which correspond with the return.

PROBLEM XIV.

To find the Form or Curvature of the Return in an open or broken
Pediment.

Let A B C be the angle which the pediment makes with the cor-

FIG. 12.

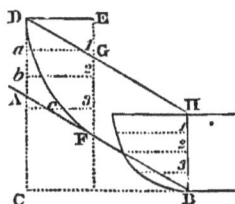

nice, and let the form and size of the moulding be as in the last problem, and as shown at D A B H.   From D drop a perpendicular on
c B, and draw D E perpendicular to D c, or parallel to c B; and let
D E be equal to E I (Fig. 11).   Then from E draw E F, parallel to
D A, and divide E F into the same number of parts as I K (Fig 11),
at 1 *a*, 2 *b*, 3 *c*, and transfer the distances 1 *a*, 2 *b*, 3 *c*, as in Fig. 11.
Then a curve line drawn through the points *a*, *b*, *c*, will be the form
of the return for the moulding of the open pediment.

The mitre for the return is cut in the usual manner, but that of
the pediment is cut to the proper angle of its inclination, as in the
last problem.   In fixing the mitre, the portion E D G of the return
must be cut away, to make it come flush with the top of the pediment
moulding.

EPITOME OF MENSURATION

AND

INSTRUMENTAL ARITHMETIC.

11

EPITOME OF MENSURATION.

OF THE CIRCLE, CYLINDER, SPHERE, &c.

1. The circle contains a greater area than any other plane figure bounded by an equal perimeter or outline.

2. The areas of circles are to each other as the squares of their diameters.

3. The diameter of a circle being 1, its circumference equals 3.1416.

4. The diameter of a circle is equal to .31831 of its circumference.

5. The square of the diameter of a circle being 1, its area equals .7854.

6 The square root of the area of a circle, multiplied by 1.12837, equals its diameter.

7. The diameter of a circle multiplied by .8862, or the circumference multiplied by .2821, equals the side of a square of equal area.

8. The sum of the squares of half the chord and versed sine divided by the versed sine, the quotient equals the diameter of corresponding circle.

9. The chord of the whole arc of a circle taken from eight times the chord of half the arc, one-third of the remainder equals the length of the arc ; or,

10. The number of degrees contained in the arc of a circle, multiplied by the diameter of the circle and by .008727, the product equals the length of the arc in equal terms of unity.

11. The length of the arc of a sector of a circle multiplied by its radius, equals twice the area of the sector.

12. The area of the segment of a circle equals the area of the sector, minus the area of a triangle whose vertex is the centre, and whose base equals the chord of the segment, or,

13. The area of a segment may be obtained by dividing the height of the segment by the diameter of the circle, and multiplying the corresponding tabular area by the square of the diameter.

14. The sum of the diameters of two concentric circles multiplied by their difference and by .7854, equals the area of the ring or space contained between them.

15. The sum of the thickness and internal diameter of a cylindric ring, multiplied by the square of its thickness and by 2.4674, equals its solidity.

16. The circumference of a cylinder, multiplied by its length or height, equals its convex surface.

17. The area of the end of a cylinder, multiplied by its length, equals its solid contents.

18. The area of the internal diameter of a cylinder, multiplied by its depth, equals its cubical capacity.

19. The square of the diameter of a cylinder multiplied by its length and divided by any other required length, the square root of the quotient equals the diameter of the other cylinder of equal contents or capacity.

20. The square of the diameter of a sphere, multiplied by 3.1416, equals its convex surface.

21. The cube of the diameter of a sphere, multiplied by .5236, equals its solid contents.

22. The height of any spherical segment or zone multiplied by the diameter of the sphere of which it is a part, and by 3.1416, equals the area or convex surface of the segment; or,

23. The height of the segment, multiplied by the circumference of the sphere of which it is a part, equals the area.

24. The solidity of any spherical segment is equal to three times the square of the radius of its base, plus the square of its height, and multiplied by its height and by 5236.

25. The solidity of a spherical zone equals the sum of the squares of the radii of its two ends, and one-third the square of its height, multiplied by the height, and by 1.5708.

26. The capacity of a cylinder, 1 foot in diameter and 1 foot in length, equals 5.875 of a United States gallon.

27. The capacity of a cylinder 1 inch in diameter and 1 foot in length, equals .0408 of a United States gallon.

28. The capacity of a cylinder, 1 inch in diameter and 1 inch in length, equals .0034 of a United States gallon.

29. The capacity of a sphere 1 foot in diameter equals 3.9156 United States gallons.

30. The capacity of a sphere 1 inch in diameter equals .002165 of a United States gallon :—hence,

31. The capacity of any other cylinder in United States gallons is obtained by multiplying the square of its diameter by its length, or the capacity of any other sphere by the cube of its diameter, and by the number of United States gallons contained as above in the unity of its measurement.

OF THE SQUARE, RECTANGLE, CUBE, &c.

1. The side of a square equals the square root of its area.

2. The area of a square equals the square of one of its sides.

3. The diagonal of a square equals the square root of twice the square of its side.

4. The side of a square is equal to the square root of half the square of its diagonal.

5. The side of a square equal to the diagonal of a given square contains double the area of the given square.

6. The area of a rectangle equals its length multiplied by its breadth.

7. The length of a rectangle equals the area divided by the breadth; or, the breadth equals the area divided by the length.

8. The side or end of a rectangle equals the square root of the sum of the diagonal and opposite side to that required, multiplied by their difference.

9. The diagonal in a rectangle equals the square root of the sum of the squares of the base and perpendicular.

10. The solidity of a cube equals the area of one of its sides multiplied by the length or breadth of one of its sides.

11. The length or breadth of a side of a cube equals the cube root of its solidity.

12. The capacity of a 12-inch cube equals 7.4784 United States gallons.

SURFACES AND SOLIDITIES OF THE REGULAR BODIES, EACH OF WHOSE
BOUNDARY LINES IS 1.

No. of sides.	Names.	Surfaces.	Solids.
4	Tetrahedron	1.7321	0.1179
6	Hexahedron	6.	1.
8	Octahedron	3.4641	0.4714
12	Dodecahedron	20.6458	7.6631
20	Icosahedron	8.6603	2.1817

The tabular surface multiplied by the square of one of the boundary lines equals the surface required ; or,

The tabular solidity multiplied by the cube of one of the boundary lines equals the solidity required.

OF TRIANGLES, POLYGONS, &c.

1. The complement of an angle is its defect from a right angle.

2. The supplement of an angle is its defect from two right angles.

3. The sine, tangent, and secant of an angle, are the cosine, cotangent, and cosecant of the complement of that angle.

4. The hypotenuse of a right-angled triangle being made radii, its sides become the sines of the opposite angles, or the cosines of the adjacent angles.

5. The three angles of every triangle are equal to two right angles: hence the oblique angles of a right-angled triangle are each others complements.

6. The sum of the squares of the two given sides of a right-angled triangle is equal to the square of the hypotenuse.

7. The difference between the squares of the hypotenuse and given side of a right-angled triangle is equal to the square of the required side.

8. The area of a triangle equals half the product of the base multiplied by the perpendicular height ; or,

9. The area of a triangle equals half the product of the two sides and the natural sine of the contained angle.

10. The side of any regular polygon multiplied by its apothem or perpendicular, and by the number of its sides, equals twice the area.

TABLE OF THE AREAS OF REGULAR POLYGONS EACH OF WHOSE
SIDES IS UNITY.

Name of Polygon.	No of Sides	Apothem or Perpend'lar	Area when Side is Unity	Interior Angle.		Central Angle.	
Triangle	3	0.2887	0.4330	60°	0	120°	0'
Square	4	0.5	1.	90	0	90	0
Pentagon	5	0.6882	1.7205	108	0	72	0
Hexagon	6	0.8660	2.5981	120	0	60	0
Heptagon	7	1.0386	3.6339	128	$34\frac{2}{7}$	51	$25\frac{2}{7}$
Octagon	8	1.2071	4.8284	135	0	45	0
Nonagon	9	1.3737	6.1818	140	0	40	0
Decagon	10	1.5388	7.6942	144	0	36	0
Undecagon	11	1.7028	9.3656	147	$16\frac{4}{11}$	32	$43\frac{7}{11}$
Dodecagon	12	1.8660	11.1962	150	0	30	0

The tabular area of the corresponding polygon multiplied by the square of the side of the given polygon equals the area of the given polygon.

OF ELLIPSES, CONES, FRUSTUMS, &c.

1. The square root of half the sum of the squares of the two diameters of an ellipse multiplied by 3.1416 equals its circumference.

2. The product of the two axes of an ellipse multiplied by .7854 equals its area.

3. The curve surface of a cone is equal to half the product of the circumference of its base multiplied by its slant side, to which, if the area of the base be added, the sum is the whole surface.

4. The solidity of a cone equals one third of the product of its base multiplied by its altitude or height.

5. The squares of the diameters of the two ends of the frustum of a cone added to the product of the two diameters, and that sum multiplied by its height and by .2618, equals its solidity.

INSTRUMENTAL ARITHMETIC,

OR UTILITY OF THE SLIDE RULE.

THE slide rule is an instrument by which the greater portion of operations in arithmetic and mensuration may be advantageously performed, provided the lines of division and gauge points be made properly correct, and their several values familiarly understood.

The lines of division are distinguished by the letters A B C D; A B and C being each divided alike, and containing what is termed a double radius,

11*

or double series of logarithmic numbers, each series being supposed to be divided into 1000 equal parts, and distributed along the radius in the following manner:

From 1 to 2 contains 301 of those parts, being the log. of 2.
```
   "     3    "    477        "           3.
   "     4    "    602        "           4.
   "     5    "    699        "           5.
   "     6    "    778        "           6.
   "     7    "    845        "           7.
   "     8    "    903        "           8.
   "     9    "    954        "           9.
```
1000 being the whole number.

The line D on the improved rules consists of only a single radius; and although of larger radius, the logarithmic series is the same, and disposed of along the line in a similar proportion, forming exactly a line of square roots to the numbers on the lines B C.

NUMERATION.

Numeration teaches us to estimate or properly value the numbers and divisions on the rule in an arithmetical form.

Their values are all entirely governed by the value set upon the first figure, and being decimally reckoned, advance tenfold from the commencement to the termination of each radius: thus, suppose 1 at the joint be one, the 1 in the middle of the rule is ten, and 1 at the end, one hundred: again, suppose 1 at the joint ten, 1 in the middle is 100, and 1 or 10 at the end is 1000, &c., the intermediate divisions on which complete the whole system of its notation.

TO MULTIPLY NUMBERS BY THE RULE.

Set 1 on D opposite to the multiplier on A; and against the number to be multiplied on B is the product on A.

Multiply 6 by 4.

Set 1 on B to 4 on A; and against 6 on B is 24 on A.

The slide thus set, against 7 on B is 28 on A.
```
                        8    "    32    "
                        9    "    36    "
                       10    "    40    "
                       12    "    48    "
                       15    "    60    "
                       25    "   100    " &c.
```

TO DIVIDE NUMBERS UPON THE RULE.

Set the divisor on B to 1 on A; and against the number to be divided on B is the quotient on A.

Divide 63 by 3.

Set 3 on B to 1 on A; and against 63 on B is 21 on A.

PROPORTION, OR RULE OF THREE DIRECT

RULE.—Set the first term on B to the second on A; and against the third upon B is the fourth upon A.

1. If 4 yards of cloth cost 38 cents, what will 30 yards cost at the same rate?

Set 4 on B to 38 on A; and against 30 on B is 285 cents on A.

2. Suppose I pay 31 dollars 50 cents for 3 cwt. of copper, at what rate is that per ton? 1 *ton* = 20 *cwt.*

Set 3 upon B to 31.5 upon A; and against 20 upon B is 210 upon A.

RULE OF THREE INVERSE.

RULE.—Invert the slide, and the operation is the same as direct proportion.

1. I know that six men are capable of performing a certain given portion of work in eight days, but I want the same performed in three; how many men must there be employed?

Set 6 upon C to 8 upon A; and against 3 upon C is 16 upon A.

2. The lever of a safety-valve is 20 inches in length, and 5 inches between the fixed end and centre of the valve; what weight must there be placed on the end of the lever to equipoise a force or pressure of 40 lbs. tending to raise the valve?

Set 5 upon C to 40 upon A; and against 20 upon C is 10 upon A.

3. If $8\frac{3}{4}$ yards of cloth, $1\frac{1}{2}$ yard in width, be a sufficient quantity, how much will be required of that which is only 7-8ths in width, to effect the same purpose?

Set 1.5 upon C to 8.75 upon A; and against .875 upon C is 15 yards upon A.

SQUARE AND CUBE ROOTS OF NUMBERS.

On the engineer's rule, when the lines C and D are equal at both ends, C is a table of squares, and D a table of roots, as

Squares	1	4	9	16	25	36	49	64	81	on C.
Roots	1	2	3	4	5	6	7	8	9	on D.

To find the geometrical mean proportion between two numbers.

Set one of the numbers upon C to the same number upon D; and against the other number upon C is the mean number or side of an equal square upon D.

Required the mean proportion between 20 and 45.

Set 20 upon C to 20 upon D; and against 45 upon C is 30 upon D.

To cube any number, set the number upon C to 1 or 10 upon D; and against the same number upon D is the cube number upon C.

Required the cube of 4.

Set 4 upon c to 1 or 10 upon D; and against 4 upon D is 64 upon c.

To extract the cube root of any number, invert the slide, and set the number upon B to 1 or 10 upon D; and where two numbers of equal value coincide on the lines B D, is the root of the given number.

Required the cube root of 64.

Set 64 upon B to 1 or 10 upon D; and against 4 upon B is 4 upon D, or root of the given number.

On the common rule, when 1 in the middle of the line c is set opposite to 10 on D, then c is a table of squares, and D a table of roots.

To cube any number by this rule, set the number upon c to 10 upon D· and against the same number upon D is the cube upon c.

MENSURATION OF SURFACE.

1. *Squares, Rectangles, &c.*

RULE.—When the length is given in feet and the breadth in inches, set the breadth on B to 12 on A; and against the length on A is the content in square feet on B.

If the dimensions are all inches, set the breadth on B to 144 upon A; and against the length upon A is the number of square feet on B.

Required the content of a board 15 inches broad and 14 feet long.

Set 15 upon B to 12 upon A; and against 14 upon A is 17.5 square feet on B.

2. *Circles, Polygons, &c.*

RULE.—Set .7854 upon c to 1 or 10 upon D; then will the lines c and D be a table of areas and diameters.

Areas 3.14 7.06 12.56 19.63 28.27 38.48 50.26 63.61 upon c.
Diam. 2   3   4   5   6   7   8   9   upon D.

In the common rule, set .7854 on c to 10 on D; then c is a line or table of areas, and D of diameters, as before.

Set 7 upon B to 22 upon A; then B and A form or become a table of diameters and circumferences of circles.

Cir. 3.14 6 28 9.42 12.56 15.7 18.85 22 25.13 28.27 upon A.
Dia. 1   2   3   4   5   6   7 8   9   upon B.

Polygons from 3 to 12 sides.—Set the gauge-point upon c to 1 or 10 upon D; and against the length of one side upon D is the area upon c.

Sides   3   5   6   7   8   9   10   11   12
Gauge-points .433 1.7 2.6 3.63 4.82 6.18 7.69 9.37 11.17

Required the area of an equilateral triangle, each side 12 inches in length.

Set .433 upon c to 1 upon D; and against 12 upon D are 62.5 square inches upon c.

TABLE OF GAUGE-POINTS FOR THE ENGINEER'S RULE.

Names.	F, F, F.	F, 1, 1.	1, 1, 1.	F, 1.	1, 1.	F.	1.
Cubic inches	578	83	1728	106	1273	105	121
Cubic feet	1	144	1	1833	22	121	33
Imp. Gallons	163	231	277	294	353	306	529
Water in lbs.	16	23	276	293	352	305	528
Gold "	814	1175	141	149	178	155	269
Silver "	15	216	261	276	334	286	5
Mercury "	118	169	203	216	258	225	389
Brass "	193	177	333	354	424	369	637
Copper "	18	26	319	331	397	345	596
Lead "	141	203	243	258	31	27	465
Wrot iron "	207	297	357	338	453	394	682
Cast iron "	222	32	384	407	489	424	733
Tin "	219	315	378	401	481	419	728
Steel "	202	292	352	372	448	385	671
Coal "	127	183	22	33	28	242	42
Marble "	591	85	102	116	13	113	195
Freestone "	632	915	11	1162	14	141	21

FOR THE COMMON SLIDE RULE.

Names.	F, F, F.	F, 1, 1.	1, 1, 1.	F, 1.	1, 1.	F.	1
Cubic inches	36	518	624	660	799	625	113
Cubic feet	625	9	108	114	138	119	206
Water in lbs.	10	144	174	184	22	191	329
Gold "	507	735	88	96	118	939	180
Silver "	938	136	157	173	208	173	354
Mercury "	738	122	127	132	162	141	242
Brass "	12	174	207	221	265	23	397
Copper "	112	163	196	207	247	214	371
Lead "	880	126	152	162	194	169	289
Wrot iron "	129	186	222	235	283	247	423
Cast iron "	139	2	241	254	304	265	458
Tin "	137	135	235	25	300	261	454
Steel "	136	183	22	233	278	239	418
Coal "	795	114	138	146	176	151	262
Marble "	370	53	637	725	81	72	121
Freestone "	394	57	69	728	873	755	132

MENSURATION OF SOLIDITY AND CAPACITY.

General Rule.—Set the length upon B to the gauge point upon A; and against the side of the square, or diameter on D, are the cubic contents, or weight in lbs. on C.

1. Required the cubic contents of a tree 30 feet in length, and 10 inches quarter girt.

Set 30 upon B to 144 (the gauge-point) upon A; and against 10 upon D is 20.75 feet upon C.

2. In a cylinder 9 inches in length, and 7 inches diameter, how many cubic inches ?

Set 9 upon B to 1273 (the gauge-point) upon A ; and against 7 on D is 346 inches on c.

3. What is the weight of a bar of cast iron 3 in. square, and 6 ft. long ?

Set 6 upon B to 32 (the gauge-point) upon A ; and against 3 upon D is 168 pounds upon c.

By the common rule.

4. Required the weight of a cylinder of wrought iron 10 inches long, and 5½ diameter.

Set 10 upon B to 283 (the gauge-point) upon A ; and against 5½ upon D is 66.65 pounds on c.

5. What is the weight of a dry rope 25 yards long, and 4 inches circumference ?

Set 25 upon B to 47 (the gauge-point) upon A ; and against 4 on D is 53 16 pounds on c.

6. What is the weight of a short-linked chain 30 yards in length, and 6-16ths of an inch in diameter ?

Set 30 upon B to 52 (the gauge-point) upon A ; and against 6 on D is 129.5 pounds on c.

POWER OF STEAM ENGINES.

Condensing Engines.—RULE. Set 3.5 on c to 10 on D ; then D is a line of diameters for cylinders, and c the corresponding number of horses' power ; thus,

| II. Pr. 3¼ | 4 | 5 | 6 | 8 | 10 | 12 | 16 | 20 | 25 | 30 | 40 | 50 | on c. |
| C. D. 10 in. | 10¾ | 12 | 13½ | 15½ | 17 | 18¾ | 21¼ | 24 | 26¾ | 29½ | 33¾ | 37¾ | on D. |

The same is effected on the common rule by setting 5 on c to 12 on D.

Non-condensing Engines.—RULE. Set the pressure of steam in pounds per square inch on B to 4 upon A ; and against the cylinder's diameter on D is the number of horses' power upon c.

Required the power of an engine, when the cylinder is 20 inches diameter and steam 30 pounds per square inch.

Set 30 on B to 4 on A ; and against 20 on D is 30 horses' power on c.

•The same is effected on the common rule by setting the force of the steam on B to 250 on A.

OF ENGINE BOILERS.

How many superficial feet are contained in a boiler 23 feet in length and 5½ feet in width ?

Set 1 on B to 23 on A ; and against 5.5 upon B is 126.5 square feet upon A.

If 5 square feet of boiler surface be sufficient for each horse-power, how many horses' power of engine is the boiler equal to ?

Set 5 upon B to 126.5 upon A ; and against 1 upon B is 25.5 upon A.

RULES AND TABLES

ARTIFICERS AND ENGINEERS.

ARTIFICERS' RULES AND TABLES

For Computing the Work of Bricklayers, Well Diggers, Masons, Carpenters and Joiners, Slaters, Plasterers, Painters, Glaziers, Pavers, and Plumbers.

MEASUREMENT OF BRICKLAYERS' WORK.

Brickwork is estimated at the rate of a number of bricks in thickness, estimating a brick at 4 inches thick. The dimensions of a building are usually taken by measuring half round on the outside, and half round on the inside; the sum of these two gives the compass of the wall,—to be multiplied by the height, for the content of the materials. Chimneys are by some measured as if they were solid, deducting only the vacuity from the hearth to the mantel, on account of the trouble of them. And by others they are girt or measured round for their breadth, and the height of the story is their height, taking the depth of the jambs for their thickness. And in this case, no deduction is made for the vacuity from the floor to the mantel-tree, because of the gathering of the breast and wings, to make room for the hearth in the next story. To measure the chimney shafts, which appear above the building, gird them about with a line for the breadth, to multiply by their height. And account their thickness half a brick more than it really is, in consideration of the plastering and scaffolding. All windows, doors, &c., are to be deducted out of the contents of the walls in which they are placed. But this deduction is made only with regard to materials; for the whole measure is taken for workmanship, and that all outside measure too, namely, measuring quite round the outside of the building, being in consideration of the trouble of the returns or angles. There are also some other allowances, such as double measure for feathered gable ends, &c.

EXAMPLE.—The end wall of a house is 28 feet long, and 37 feet high to the eaves: 15 feet high is four bricks or 16 inches thick, other 12 feet is three bricks or 12 inches thick, and the remaining 10 feet is two bricks or 8 inches thick; above which is a triangular gable 12 feet high and one brick or 4 inches in thickness. What number of bricks are there in the said wall? *Ans.* 25,620.

```
                          thickness.
    28 × 15 = 420 × 4 = 1680 contents of 1st story.
    28 × 12 = 336 × 3 = 1008    "     "  2d    "
    28 × 10 = 280 × 2 =  560    "     "  3d    "
12 ÷ 2 =  6 × 28 = 168 × 1 =  168    "     "  gable.

               3416  square feet area of whole wall.
                   7½ bricks to square foot.

               23,912    By the table
                1,708    3000 suprfi. ft. = 22,500 bricks.
               ------     400  "      "  =  3,000   "
Answer,—      25,620 bricks.   10  "      "  =     75   "
                                6  "      "  =     45   "
                                   ----
                               3416   "      "  = 25,620 bricks
```

A Table by which to ascertain the number of Bricks necessary to construct any Piece of Building, from a four-inch Wall to twenty-four inches in Thickness.

The utility of the Table (on next page) can be seen by the following Example. Required the number of bricks to build a wall of 12 inches thickness, and containing an area of 6,437 square feet.

```
Square feet 1000      22,500 bricks—See table.
         × 6             6
        ----            ----
       6000 =   135.000           NOTE.—7½ bricks,
        400 =     9,000      equal one superficial foot.
         30 =       675
          7 =       153
        ----      -------
      6,437 =   144,833 bricks.
```

Superficial feet of Wall.	Number of Bricks to Thickness of					
	4-inch.	8-inch.	12-inch.	16-inch.	20-inch.	24-inch.
1	8	15	23	30	38	45
2	15	30	45	60	75	90
3	23	45	68	90	113	135
4	30	60	90	120	150	180
5	38	75	113	150	188	225
6	45	90	135	180	225	270
7	53	105	158	210	263	315
8	60	120	180	240	300	360
9	68	135	203	270	338	405
10	75	150	225	300	375	450
20	150	300	450	600	750	900
30	225	450	675	900	1125	1350
40	300	600	900	1200	1500	1800
50	375	750	1125	1500	1875	2250
60	450	900	1350	1800	2250	2700
70	525	1050	1575	2100	2625	3150
80	600	1200	1800	2400	3000	3600
90	675	1350	2025	2700	3375	4050
100	750	1500	2250	3000	3750	4500
200	1500	3000	4500	6000	7500	9000
300	2250	4500	6750	9000	11250	13500
400	3000	6000	9000	12000	15000	18000
500	3750	7500	11250	15000	18750	22500
600	4500	9000	13500	18000	22500	27000
700	5250	10500	15750	21000	26250	31500
800	6000	12000	18000	24000	30000	36000
900	6750	13500	20250	27000	33750	40500
1000	7500	15000	22500	30000	37500	45000

MEASUREMENT OF WELLS AND CISTERNS.

There are two methods of estimating the value of excavating. It may be done by allowing so much a day for every man's work, or so much per cubic foot, or yard, for all that is excavated.

Well Digging. — Suppose a Well is 40 feet deep, and 5 feet in diameter, required the number of cubic feet, or yards?

$$5 \times 5 = 25 \times .7854 = 19.635 \times 40 = 785.4 \text{ cubic feet.}$$

Suppose a well is he 4 feet 9 inches diameter, and $16\frac{1}{2}$ feet from the bottom to the surface of the water ; how many gallons are therein contained ?

$$4.75^2 \times 16.5 \times 5.875 = 2187.152 \text{ gallons.}$$

Again, suppose the well's diameter the same, and its entire depth 35 feet ; required the quantity in cubic yards of material excavated in its formation.

$$4.75^2 \times 35 \times .02909 = 22.972 \text{ cubic yards.}$$

A cylindrical piece of lead is required $7\frac{1}{2}$ inches diameter, and 168 lbs. in weight; what must be its length in inches ?

$$7.5^2 \times .3223 = 18, \text{ and } 168 \div 18 = 9.3 \text{ inches.}$$

Digging for Foundations, &c. — To find the cubical quantity in a trench, or an excavated area, the length, width, and depth must be multiplied together. These are usually given in feet, and therefore, to reduce the amount into cubic yards it must be divided by 27.

Suppose a trench is 40 feet long, 3 feet wide, and 3 feet deep, required the number of cubic feet, or yards ?

$$40 \times 3 = 120 \times 3 = 360 \text{ feet} \div 27 = 13\frac{1}{3} \text{ yards.}$$

24 cubic feet of sand, 17 ditto clay, 18 ditto earth, equal one ton.

1 cubic yard of earth or gravel, before digging, will occupy about $1\frac{1}{2}$ cubic yards when dug.

MEASUREMENT OF MASONS' WORK.

To masonry belong all sorts of stone-work ; and the measure made use of is a foot, either superficial or solid.

Walls, columns, blocks of stone or marble, &c., are measured by the cubic

12

foot; and pavements, slabs, chimney-pieces, &c., by the superficial or square foot. Cubic or solid measure is used for the materials, and square measure for the workmanship. In the solid measure, the true length, breadth and thickness, are taken, and multiplied continually together. In the superficial, there must be taken the length and breadth of every part of the projection, which is seen without the general upright face of the building.

EXAMPLE.— In a chimney-piece, suppose the length of the mantel and slab each 4 feet 6 inches ; breadth of both together 3 feet 2 inches; length of each jamb 4 feet 4 inches; breadth of both together 1 foot 9 inches. Required the superficial content. — *Ans.* 21 feet 10 inches.

$$\left. \begin{array}{l} 4 \text{ ft. 6 in.} \times 3 \text{ ft. 2 in.} = 14 \text{ ft. 3 in.} \\ 4 \text{ " 4 "} \times 1 \text{ " 9 "} = 7 \text{ " 7 "} \end{array} \right\} 21 \text{ feet 10 inches.}$$

Rubble Walls (unhewn stone) are commonly measured by the perch, which is 16½ feet long, 1 foot deep, and 1½ foot thick, equivalent to 24¾ cubic feet. 25 cubic feet is sometimes allowed to the perch, in measuring stone before it is laid, and 22 after it is laid in the wall. This species of work is of two kinds, coursed and uncoursed ; in the former the stones are gauged and dressed by the hammer, and the masonry laid in horizontal courses, but not necessarily confined to the same height. The uncoursed rubble wall is formed by laying the stones in the wall as they come to hand, without any previous gauging or working.

27 cubic feet of mortar require for its preparation, 9 bushels of lime and 1 cubic foot of sand.

Lime and sand lessen about one-third in bulk when made into mortar ; likewise cement and sand.

Lime, or cement and sand, to make mortar, require as much water as is equal to one-third of their bulk.

All sandstones ought to be placed on their natural beds ; from inattention to this circumstance, the stones often split off at the joints, and the position of the lamina much sooner admits of the destructive action of air and water.

The heaviest stones are most suited for docks and harbors, breakwaters to bridges, &c.

Granite is the most durable species of stone yet known for the purposes of building. It varies in weight according to quality ; the heaviest is the most durable.

MEASUREMENT OF CARPENTERS' AND JOINERS' WORK.

To this branch belongs all the wood work of a house, such as flooring, partitioning, roofing, &c. Large and plain articles are usually measured by the square foot or yard, &c., but enriched mouldings, and some other articles, are often estimated by running or lineal measures, and some things are rated by the piece.

All *joints, girders,* and in fact all the parts of naked flooring, are measured by the cube, and their quantities are found by multiplying the length by the breadth, and the product by the depth. The same rule applies to the measurement of all the timbers of a roof, and also the framed timbers used in the construction of partitions.

Flooring, that is to say, the boards which cover the naked flooring, is measured by the square. The dimensions are taken from wall to wall, and the product is divided by 100, which gives the number of squares ; but deductions must be made for staircases and chimneys.

In measuring of joists, it is to be observed, that only one of their dimensions is the same with that of the floor ; for the other exceeds the length of the room by the thickness of the wall, and one-third of the same, because each end is let into the wall about two-thirds of its thickness.

No deductions are made for hearths, on account of the additional trouble and waste of materials.

Partitions are measured from wall to wall for one dimension, and from floor to floor, as far as they extend, for the other.

No deduction is made for door-ways, on account of the trouble of framing them.

In measuring of joiners' work, the string is made to ply close to every part of the work over which it passes.

The measure for centering for CELLARS is found by making a string pass over the surface of the arch for the breadth, and taking the length of the cellar for the length; but in groin centering, it is usual to allow double measure, on account of their extraordinary trouble.

In *roofing*, the length of the house in the inside, together with two-thirds of the thickness of one gable, is to be considered as the length, and the breadth is equal to double the length of a string which is stretched from the ridge down the rafter, and along the eaves-board, till it meets with the top of the wall.

For *staircases*, take the breadth of all the steps, by making a line ply close over them, from the top to the bottom, and multiply the length of this line by the length of a step, for the whole area.— By the length of a step is meant the length of the front and the returns at the two ends; and by the breadth, is to be understood the girth of its two outer surfaces, or the tread and riser.

For the *balustrade*, take the whole length of the upper part of the handrail, and girt over its end till it meet the top of the newel post, for the length; and twice the length of the baluster upon the landing, with the girth of the handrail for the breadth.

For *wainscoting*, take the compass of the room for the length; and the height from the floor to the ceiling, making the string ply close into all the mouldings for the breadth. Out of this must be made deductions for windows, doors, and chimneys, &c., but workmanship is counted for the whole, on account of the extraordinary trouble.

For *doors*, it is usual to allow for their thickness, by adding it to both dimensions of length and breadth, and then to multiply them together for the area. If the door be paneled on both sides, take double its measure for the workmanship; but if the one side only be paneled, take the area and its half for the workmanship. — For the *surrounding architrave*, gird it about the outermost parts for its length; and measure over it, as far as it can be seen when the door is open, for the breadth.

Window-shutters, bases, &c., are measured in the same manner.

In the measuring of roofing for workmanship alone, holes for chimney-shafts and sky-lights are generally deducted. But in measuring for work and materials, they commonly measure in all sky-lights, lutheran lights, and holes for the chimney-shafts, on account of their trouble and waste of materials.

The doors and shutters, being worked on both sides, are reckoned work and half work.

Hemlock and *Pine* Shingles are generally 18 inches long, and of the average width of 4 inches. When nailed to the roof 6 inches are generally left out to the weather, and 6 shingles are therefore required to a square foot. *Cedar* and *Cypress* Shingles are generally 20 inches long, and 6 inches wide, and therefore a less number are required for a "square." On account of waste and defects, 1000 shingles should be allowed to a square.

Two 4-penny nails are allowed to each shingle, equal to 1200 to a square.

The weight of a square of partitioning may be estimated at from 1500 to 2000 lbs.; a square of single-joisted flooring, at from 1200 to 2000 lbs.; a square of framed flooring, at from 2700 to 4500 lbs; a square of deafening, at about 1500 lbs. 100 superficial feet make one square of boarding, flooring, &c.

In selecting Timber, avoid spongy heart, porous grain, and dead knots; choose the brightest in color, and where the strong red grain appears to rise on the surface.

The *Carpenter* will find in the "Business Man's Assistant" TABLES giving the solid contents of Timber and Logs; the square feet in Scantling from 2.2 to 15.16 inches; the square feet in Boards and Planks; the contents of Logs in standard Board measure; the strength and weight of Iron Cylinders, Trusses, Plates, Cast Iron for Beams, and Hoop Iron.

Number of American Iron Machine Cut Nails, in a pound, (by count.)

Size.	Number.	Size.	Number.	Size.	Number.
3 penny	. . 408	6 penny	. . 156	12 penny	. . . 52
4 "	. . . 275	8 "	. . . 100	20 "	 32
5 "	. . . 227	10 "	. . . 66	30 "	 25

SASH TABLE.— *Size* and *Prices* of *Sashes*, *Shutters*, *&c.* Cincinnati, Ohio.

Size of Lights.	Thickness.	Size of Sash for 12 light Windows.		Price of Sash per Light.	Price of Venitian Shutters per pair.	Price of Window Frames.	
		Width.	Length.			Box.	Common.
Inches.	In.	feet. in.	feet. in.	cts.	$ cts.	$ cts.	$ cts.
8 by 10	1⅛	2 4	3 10	4	1 37½	2 00	1 20
8 by 10	1⅜	2 4	3 10	5	1 62½	2 00	1 20
9 by 12	1⅛	2 7½	4 6½	5	1 62½	2 50	1 30
9 by 12	1⅜	2 7½	4 6½	6	1 75	2 50	1 30
10 by 12	1⅛	2 10½	4 6½	5	1 62½	2 50	1 30
10 by 12	1⅜	2 10½	4 6½	6	1 75	2 50	1 30
10 by 14	1⅜	2 10½	5 2½	7	2 12½	2 75	1 40
10 by 15	1⅜	2 10½	5 6½	7½	2 25	2 75	1 40
10 by 16	1⅜	2 10½	5 10½	8	2 37½	3 20	1 50
11 by 15	1⅜	3 2	5 6½	8	2 37½	3 20	1 50
11 by 16	1⅜	3 2	5 10½	8½	2 50	3 35	1 60
11 by 17	1⅜	3 2	6 2½	8½	2 62½	3 50	1 70
12 by 16	1⅜	3 5	5 10½	8½	2 62½	3 75	1 80
12 by 18	1⅜	3 5	6 6½	9	2 87½	4 00	1 90
12 by 20	1⅜	3 5	7 2½	10	3 12½	4 25	2 12½
12 by 22	1⅜	3 5	7 10½	11	3 37½	4 50	2 30
12 by 24	1⅜.	3 5	8 6½	12	3 62½	4 75	2 50

Sash 1 1-2 or 1 3-4 inches thick, add 1 1-2 cents per light, to 1 3-8 inch prices; for Ploughing and Boring sash, add 1-2 cent per light; all 1 3-8 sash are made with hook rails.

Venitian Shutters, 1 1-2 or 1 3-4 inches thick, add 50 cents per pair to 1 3-8 inch prices. Shutters are made 1 1-4 inches longer than sash. Pivot or Rolling Shutters, extra price.

MEASUREMENT OF SLATERS' WORK.

In these articles, the content of a roof is found by multiplying the length of the ridge by the girth over from eaves to eaves; making allowance in this girth for the double row of slates at the bottom, or for how much one row of slates is laid over another. When the roof is of a true pitch, that is, forming a right angle at top, then the breadth of the building with its half added, is the girth over both sides. In angles formed in a roof, running from the ridge to the eaves, when the angle bends inwards, it is called a valley; but when outwards, it is called a hip. It is not usual to make deductions for chimney-shafts, sky-lights or other openings.

SLATES. [*From the Quarries of Rutland County, Vermont.*]

3 inch Cover.		2 inch Cover.	3 inch Cover.		2 inch Cover.
Sizes of Slates.	No. of Slates to the Square or 100 Feet.	No. of slates to the square or 100 Feet.	Sizes of Slates.	No. of Slates to the Square or 100 Feet.	No. of slates to the square or 100 Feet.
24 by 16	86	84	18 by 11	174¼	163½
24 by 14	98	93½	18 by 10	192	189
24 by 12	114	109	18 by 9	213	200
22 by 14	108	102¼	16 by 12	184	171½
22 by 12	126	120	16 by 10	221½	205¾
22 by 10	152	144	16 by 9	246	228½
20 by 14	129	114¼	16 by 8	277	257
20 by 12	143	133⅓	14 by 10	262	240
20 by 11	146	145½	14 by 9	293	266½
20 by 10	169¼	160	14 by 8	327	300
18 by 12	160	150	14 by 7	374	343

"Each Slate is 3 inches BOND or COVER. The rule for measuring Slating is, to add one foot for all hips and valleys. No deduction is made for Lutheran windows, skylights or chimneys, except they are of unusual size; then one half is deducted."

IMPORTED SLATES.

Names of Slates.	Sizes.	Number of Super-ficial Feet each M of 1200 will cover.	Weight of each M of 1200 Slates.
	Inches. Inches.		
Duchesses, ...	24 by 12	1100	60 cwt.
Marchionesses, . . .	22 " 12	1000	55 "
Countesses, ...	20 " 10	750	40 "
Viscountesses, . . .	18 " 10	666 2-3	36 "
Ladies, ...	16 " 10	583 1-3	31 "
do.	16 " 8	466 2-3	25 "
do.	14 " 8	400	22 "
do.	12 " 8	333 1-3	18 1-2 "
Plantations, ...	14 " 12	600	33 "
do.	13 " 10	458 1-3	25 "
do.	12 " 10	416 2-3	23 "
Doubles, ...	13 " 7	320 5-6	17 1-2 "
do. small, . . .	11 " 7	262 1-2	14 1-2 "
School Slates for	5 ft. by 2 1-2 ft		
Blackboards, ...	5 feet by 3 feet.		

MEASUREMENT OF PLASTERERS' WORK.

Plasterers' work is of two kinds, namely, ceiling—which is plastering upon laths —and rendering, which is plastering upon walls, which are measured separately.

The contents are estimated either by the foot or yard, or square of 100 feet. Enriched mouldings, &c., are rated by running or lineal measure. One foot extra is allowed for each mitre.

One half of the openings, windows, doors, &c., allowed to compensate for trouble of finishing returns at top and sides.

Cornices and mouldings, if 12 inches or more in girt, are sometimes estimated by the sq ft. ; if less than 12 inches they are usually measured by the lineal foot.

1 bushel of cement will cover 1 1-7 square yards at 1 inch in thickness.

do.   do.   do. $1\frac{1}{2}$ do.   do. $\frac{3}{4}$ do.   do.

do.   do.   do. $2\frac{1}{4}$ do.   do. $\frac{1}{2}$ do.   do.

1 bushel of cement and 1 of sand will cover $2\frac{1}{2}$ sq. yds. at 1 inch in thickness.

do.   do.   do.   do.   3 do. $\frac{3}{4}$ do.   do.

do.   do.   do.   do. $4\frac{1}{2}$ do. $\frac{1}{2}$ do.   do.

1 bushel of cement and 2 of sand will cover $3\frac{1}{4}$ square yds. at 1 inch in thickness.

do.   do.   do.   do. $4\frac{1}{2}$ do. $\frac{3}{4}$ do.   do.

do.   do.   do.   do. $6\frac{3}{4}$ do. $\frac{1}{2}$ do.   do.

1 cwt. of mastic and 1 gallon of oil will cover $1\frac{1}{2}$ yards at $\frac{3}{4}$, or $2\frac{1}{2}$ at $\frac{1}{4}$ inch,

1 cubic yard of lime, 2 yards of road or drift sand, and 3 bushels of hair. will cover 75 yards of *render and set* on brick, and 70 yards on lath, or 65 yards *plaster*, or *render*, 2 *coats and set* on brick, and 60 yards on lath ; floated work will require about the same as 2 coats and set.

Laths are $1\frac{1}{4}$ to $1\frac{1}{2}$ inches by 4 feet in length, and are usually set $\frac{1}{4}$th of an inch apart. A bundle contains 100. 1 bundle of laths and 500 nails cover about $4\frac{1}{2}$ yds.

MEASUREMENT OF PAVERS' WORK.

Pavers' work is done by the square yard. And the content is found by multi-plying the length by the breadth. Grading for paving is charged by the day.

MEASUREMENT OF PAINTERS' WORK.

Painters' work is computed in square yards. Every part is measured where the color lies ; the measuring line is forced into all the mouldings and corners.

12*

Cornices, mouldings, narrow skirtings, reveals to doors and windows, and generally all work not more than nine inches wide, are valued by their length. Sash-frames are charged so much each according to their size, and the squares so much a dozen. Mouldings, cut in, are charged by the foot run, and the workman always receives an extra price for party-colors. Writing is charged by the inch, and the price given is regulated by the skill and manner in which the work is executed; the same is true of imitations and marbling. The price of painting varies exceedingly, some colors being more expensive and requiring much more labor than others. In measuring open railing, it is customary to take it as flat work, which pays for the extra labor; and as the rails are painted on all sides, the two surfaces are taken. It is customary to allow all edges and sinkings.

MEASUREMENT OF GLAZIERS' WORK.

Glaziers' work is sometimes measured by the sq. ft., sometimes by the piece, or at so much per light; except where the glass is set in metallic frames, when the charge is by the foot. In estimating by the sq. ft., it is customary to include the whole sash. Circular or oval windows are measured as if they were square.

TABLE SHOWING THE SIZE AND NUMBER OF LIGHTS TO THE 100 SQUARE FEET.

Size.	Lights.	Size.	Lights.	Size.	Lights.	Size.	Lights
6 by 8	300	12 by 14	86	14 by 22	47	20 by 20	36
7 by 9	229	12 by 15	80	14 by 24	43	20 by 22	33
8 by 10	180	12 by 16	75	15 by 15	64	20 by 24	30
8 by 11	164	12 by 17	71	15 by 16	60	20 by 25	29
8 by 12	150	12 by 18	67	15 by 18	53	20 by 26	28
9 by 10	160	12 by 19	63	15 by 20	48	20 by 28	26
9 by 11	146	12 by 20	60	15 by 21	46	21 by 27	25
9 by 12	133	12 by 21	57	15 by 22	44	22 by 24	27
9 by 13	123	12 by 22	55	15 by 24	40	22 by 26	25
9 by 14	114	12 by 23	52	16 by 16	56	22 by 28	23
9 by 16	100	12 by 24	50	16 by 17	53	24 by 28	21
10 by 10	144	13 by 14	79	16 by 18	50	24 by 30	20
10 by 12	120	13 by 15	74	16 by 20	45	24 by 32	19
10 by 13	111	13 by 16	69	16 by 21	43	25 by 30	19
10 by 14	103	13 by 17	65	16 by 22	41	26 by 36	15
10 by 15	96	13 by 18	61	16 by 24	38	28 by 34	15
10 by 16	90	13 by 19	58	17 by 17	50	30 by 40	12
10 by 17	85	13 by 20	55	17 by 18	47	31 by 36	13
10 by 18	80	13 by 21	53	17 by 20	42	31 by 40	12
11 by 11	119	13 by 22	50	17 by 22	38	31 by 42	12
11 by 12	109	13 by 24	46	17 by 24	35	32 by 42	10
11 by 13	101	14 by 14	73	18 by 18	44	32 by 44	10
11 by 14	94	14 by 15	68	18 by 20	40	33 by 45	10
11 by 15	87	14 by 16	64	18 by 22	36	34 by 46	9
11 by 16	82	14 by 17	60	18 by 24	33	30 by 52	9
11 by 17	77	14 by 18	57	19 by 19	40	32 by 56	8
11 by 18	73	14 by 19	54	19 by 20	38	33 by 56	8
12 by 12	100	14 by 20	51	19 by 22	34	36 by 58	7
12 by 13	92	14 by 21	49	19 by 24	32	38 by 58	7

MEASUREMENT OF PLUMBERS' WORK.

Plumbers' work is rated at so much a pound, or else by the hundred weight, of 112 pounds. Sheet lead, used in roofing, guttering, &c., is from 7 to 12 lbs. to the square foot. And a pipe of an inch bore is commonly from 6 to 13 lbs. to the yard in length. — [See Table, " *Weight of Lead Pipe per Foot*"]

PATENT IMPROVED LEAD PIPE, SIZES AND WEIGHT PER FOOT.

Calibre.	Weight per foot.	Calibre	Weight per foot.	Calibre	Weight per foot.	Calibre	Weight per foot.	Calibre.	Weight per foot.
Inches.	lbs. ozs.	Inches.	lbs. ozs.	Inches.	lbs. ozs.	Inches.	lbs. ozs.	Inches.	lbs. ozs.
⅜	6	½	1 4	¾	1 4	1	4 0	1½	5 0
"	8	"	1 8	"	2 0	"	6 0	1¾	4 0
"	10	"	2 0	"	2 4	1¼	2 8	2	5 0
"	12	"	3 0	"	2 8	"	3 0	"	6 0
"	1 0	⅝	13	"	3 0	"	3 8	"	7 0
"	1 8	"	1 0	"	4 0	"	4 0	2½ (1-4 in. thick.)	11 0
½	8	"	1 8	1	1 8	"	5 0	3	13 0
"	10	"	2 0	"	1 12	1½	3 0	3½	15 0
"	12	"	2 12	"	2 0	"	3 8	4	18 0
"	14	¾	12	"	2 8	"	4 0	4½	20 0
"	1 0	"	14	"	3 0	"	4 8	5	22 0

Sheet Lead.—Weight of a Square Foot, 2½, 3, 3½, 4, 4½, 5, 6, 7, 8½, 9, 10 lbs. and upwards.

BOSTON LEAD PIPE, SIZES AND WEIGHT PER FOOT.

1-2 Inch.		5-8 Inch.		3-4 Inch.		1 Inch.		1 1-4 Inch.		1 1-2 Inch.		1 3-4 Inch.		2 Inch.	
lbs.	*oz.*	*lbs.*	*oz.*	*lbs.*	*oz.*	*lbs.*	*oz.*	*lbs.*	*oz.*	*lbs.*	*oz.*	*lbs.*	*oz.*	*lbs.*	*oz.*
	10	2	12	1	1	1	8	2	4	3	5	3	10	4	12
	12	3		1	6	1	12	2	8	3	12	4	3	5	8
	16			1	12	2		2	13	4	4	5	2	7	12
1	4			2	4	2	6	3	3	4	10				
1	8			3	2	2	14	3	15	6					
1	11			3	14	3	13								
1	14					5									
2	4					6	4								

COMPARATIVE STRENGTH AND WEIGHT OF ROPES AND CHAINS.

Circum. of Rope in inches.	Weight per Fathom in lbs.	Diameter of Chain in inches.	Weight per Fathom in lbs.	Proof Strength in tons and cwt.	Circum. of Rope in inches.	Weight per Fathom in lbs.	Diameter of Chain in inches.	Weight per Fathom in lbs.	Proof Strength in tons and cwt.
3½	2¾	$\frac{5}{16}$	5½	1 5½	10	23	⅞	43	10 0
4¼	4¾	⅜	8	1 16¾	10¾	28	$\frac{15}{16}$	49	11 11
5	5¾	$\frac{7}{16}$	10½	2 10	11½	30½	1 in.	56	13 8
5¾	7	½	14	3 5½	12¼	36	$1\frac{1}{16}$	63	14 18
6½	9¾	$\frac{9}{16}$	18	4 3½	13	39	1⅛	71	16 14
7	11¼	⅝	22	5 2	13¾	45	$1\frac{3}{16}$	79	18 11
8	15	$\frac{11}{16}$	27	6 4½	14½	48½	1¼	87	20 8
8¾	19	¾	32	7 7	15¼	56	$1\frac{5}{16}$	96	22 13
9½	21	$\frac{13}{16}$	37	8 13½	16	60	1⅜	106	24 18

Note.—It must be understood and also borne in mind, that, in estimating the amount of tensile strain to which a body is subjected, the weight of the body itself must also be taken into account; for according to its position so may it approximate to its whole weight in tending to produce extension within itself; as in the almost constant application of ropes and chains to great depths, considerable heights, &c.

STRENGTH OF MATERIALS OF CONSTRUCTION.

[*From Templeton's Workshop Companion.*]

MATERIALS of construction are liable to four different kinds of strain; viz., stretching, crushing, transverse action, and torsion or twisting: the first of which depends upon the body's tenacity alone; the second, on its resistance to compression; the third, on its tenacity and compression combined; and the fourth, on that property by which it opposes any acting force tending to change from a straight line, to that of a spiral direction, the fibres of which the body is composed.

In bodies, the power of tenacity and resistance to compression, in the direction of their length, is as the cross section of their area multiplied by the results of experiments on similar bodies, as exhibited in the following tables.

Table showing the Tenacities, Resistances to Compression, and other Properties of the common Materials of Construction.

Names of Bodies.	Absolute.		Compared with Cast Iron.		
	Tenacity in lbs. per sq. inch.	Resistance to compression in lbs. per sq. inch.	Its strength is	Its extensibility is	Its stiffness is
Ash, . . .	14130		0.23	2.6	0.089
Beech, . . .	12225	8548	0.15	2.1	0.073
Brass, . . .	17968	10304	0.435	0.9	0.49
Brick, . . .	275	562			
Cast Iron, . .	13434	86397	1.000	1.0	1.000
Copper (wrought), .	33000				
Elm, . . .	9720	1033	0.21	2.9	0.073
Fir, or Pine, white,	12346	2028	0.23	2.4	0.1
" " Red, .	11800	5375	0.3	2.4	0.1
" " Yellow,	11835	5445	0.25	2.9	0.087
Granite (Aberdeen),		10910			
Gun-metal (copper 8, and tin 1). . .	35838		0.65	1.25	0.535
Malleable Iron, .	56000		1.12	0.86	1.3
Larch, . . .	12240	5568	0.136	2.3	0.0585
Lead, . . .	1824		0.096	2 5	0.038
Mahogany, Honduras,	11475	8000	0 24	2.9	0.487
Marble, . . .	551	6060			
Oak, . . .	11880	9504	0.25	2.8	0.093
Rope (1 in. in circum.)	200				
Steel, . . .	128000				
Stone, Bath, . .	478				
" Craigleith, ,	772	5490			
" Dundee, .	2661	6630			
" Portland, .	857	3729			
Tin (cast) . .	4736		0 182	0.75	0 25
Zinc (sheet) . .	9120		0 365	0.5	0.76

RESISTANCE TO LATERAL PRESSURE, OR TRANSVERSE ACTION.

The strength of a square or rectangular beam to resist lateral pressure, acting in a perpendicular direction to its length, is as the breadth and square of the depth, and inversely as the length;—thus, a beam twice the breadth of another, all other circumstances being alike, equal twice the strength of the other; or twice the depth, equal four times the strength, and twice the length, equal only half the strength, &c., according to the rule.

Table of Data, containing the Results of Experiments on the Elasticity and Strength of various Species of Timber, by Mr. Barlow.

Species of Timber.	Value of E.	Value of S.	Species of Timber.	Value of E.	Value of S.
Teak, . .	174.7	2462	Elm, . . .	50.64	1013
Poona, .	122.26	2221	Pitch pine, .	88.68	1632
English Oak, .	105.˙	1672	Red pine, . .	133.	1341
Canadian do. .	155.5	1766	New England Fir.	158.5	1102
Dantzic do. .	86.2	1457	Riga Fir, . .	90.	1100
Adriatic do. .	70.5	1383	Mar Forest do.	63.	1200
Ash, . . .	119.	2026	Larch, . .	76.	900
Beech, . . .	98.	1556	Norway Spruce.	105.47	1474

To find the dimensions of a beam capable of sustaining a given weight, with a given degree of deflection, when supported at both ends.

RULE.—Multiply the weight to be supported in lbs. by the cube of the length in feet; divide the product by 32 times the tabular value of E, multiplied into the given deflection in inches; and the quotient is the breadth multiplied by the cube of the depth in inches.

NOTE 1.—When the beam is intended to be square, then the fourth root of the quotient is the breadth and depth required.

∗ NOTE 2.—If the beam is to be cylindrical, multiply the quotient by 1.7, and the fourth root of the product is the diameter.

Ex. The distance between the supports of a beam of Riga fir is 16 feet, and the weight it must be capable of sustaining in the middle of its length is 8000 lbs, with a deflection of not more than ¾ of an inch; what must be the depth of the beam, supposing the breadth 8 inches?

$$\frac{16 \times 8000}{90 \times 32 \times .75} = 15175 \div 8 = \sqrt[3]{1897} = 12.35 \text{ in., the depth.}$$

To determine the absolute strength of a rectangular beam of timber, when supported at both ends, and loaded in the middle of its length, as beams in general ought to be calculated to, so that they may be rendered capable of withstanding all accidental cases of emergency.

RULE.—Multiply the tabular value of S by four times the depth of the beam in inches, and by the area of the cross section in inches; divide the product by the distance between the supports in inches, and the quotient will be the absolute strength of the beam in lbs.

NOTE 1.—If the beam be not laid horizontally, the distance between the supports, for calculation, must be the horizontal distance.

NOTE 2.—One fourth of the weight obtained by the rule, is the greatest weight that ought to be applied in practice as permanent load.

NOTE 3.—If the load is to be applied at any other point than the middle, then the strength will be as the product of the two distances is to the square of half the length of the beam between the supports;—or, twice the distance from one end, multiplied by twice from the other, and divided by the whole length, equal the effective length of the beam.

Ex. In a building 18 feet in width, an engine boiler of 5½ tons (2240 lbs. to a ton) is to be fixed, the center of which is to be 7 feet from the wall, and having two pieces of red pine, 10 inches by 6, which I can lay across the two walls for the purpose of slinging it at each end,—may I with sufficient confidence apply them, so as to effect this object?

2240×5.5 ÷ 2 = 6160 lbs. to carry at each end.

And 18 feet — 7 = 11, double each, or 14 and 22, then 14×22 ÷ 18 = 17 feet, or 204 inches, effective length of beam.

Tabular value of S, red pine, =1341×4×10×60 ÷ 204 = 15776 lbs. the absolute strength of each piece of timber at that point.

To determine the dimensions of a rectangular beam capable of supporting a required weight, with a given degree of deflection, when fixed at one end.

RULE.—Divide the weight to be supported, in lbs., by the tabular value of E, multiplied by the breadth and deflection, both in inches ; and the cube root of the quotient, multiplied by the length in feet, equal the depth required in inches.

EX. A beam of ash is intended to bear a load of 700 lbs. at its extremity ; its length being 5 feet, breadth 4 inches, and the deflection not to exceed ¼ an inch.

Tabular value of E $= 119 \times 4 \times .5 = 238$ the divisor ;

then $700 \div 238 = \sqrt[3]{2.94} \times 5 = 7.25$ inches, depth of the beam.

To find the absolute strength of a rectangular beam, when fixed at one end, and loaded at the other

RULE.—Multiply the value of S by the depth of the beam, and by the area of its section, both in inches ; divide the product by the leverage in inches, and the quotient equal the absolute strength of the beam in lbs.

EX. A beam of Riga fir, 12 inches by 4½, and projecting 6½ feet from the wall ; what is the greatest weight it will support at the extremity of its length ?

Tabular value of S $= 1100$. $12 \times 4.5 = 54$ sectional area.

Then, $1100 \times 12 \times 54 \div 78 = 9139.4$ lbs.

When fracture of a beam is produced by vertical pressure, the fibres of the lower section of fracture are separated by extension, whilst at the same time those of the upper portion are destroyed by compression ; hence exists a point in section where neither the one nor the other takes place, and which is distinguished as the point of neutral axis. Therefore, by the law of fracture thus established, and proper data of tenacity and compression given, as in the preceding table, we are enabled to form metal beams of strongest section with the least possible material. Thus, in cast iron, the resistance to compression is nearly as 6½ to 1 of tenacity, consequently a beam of cast iron, to be of strongest section, must be of the following form, and a parabola in the direction of its length, the quantity of material in the bottom flange being about 6½ times that of the upper. But such is not the case with beams of timber ; for although the tenacity of timber be on an average twice that of its resistance to compression, its flexibility is so great, that any considerable length of beam, where columns cannot be situated to its support, requires to be strengthened or trussed by iron rods, as in the following manner.

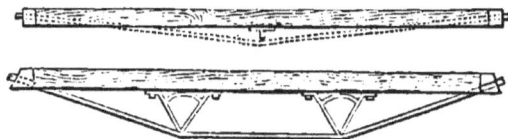

And these applications of principle not only tend to diminish deflection, but the required purpose is also more effectively attained, and that by lighter pieces of timber.

To ascertain the absolute strength of a cast iron beam of the preceding form, or that of strongest section.

RULE.—Multiply the sectional area of the bottom flange in inches by the depth of the beam in inches, and divide the product by the distance between the supports, also in inches ; and 514 times the quotient equal the absolute strength of the beam in cwts.

The strongest form in which any given quantity of matter can be disposed is that of a hollow cylinder ; and it has been demonstrated that the maximum of strength is obtained in cast iron, when the thickness of the annulus, or ring, amounts to one-fifth of the cylinder's external diameter ; the relative strength of a solid to that of a hollow cylinder being as the diameters of their sections. (*See Tables.*)

A Table showing the Weight or Pressure a beam of Cast Iron, 1 inch in breadth, will sustain, without destroying its elastic force, when it is supported at each end, and loaded in the middle of its length, and also the deflection in the middle which that weight will produce. By Mr. Hodgkinson, Manchester.

Length.	6 feet.		7 feet.		8 feet.		9 feet.		10 feet.	
Depth in in.	Weight in lbs.	Defl. in in.	Weight in lbs.	Defl. in in.	Weight in lbs.	Defl. in in.	Weight in lbs.	Defl. in in.	Weight in lbs.	Defl. in in
3	1278	.24	1089	.33	954	.426	855	.54	765	.66
3½	1739	.205	1482	.28	1298	.365	1164	.46	1041	.57
4	2272	.18	1936	.245	1700	.32	1520	.405	1360	.5
4½	2875	.16	2450	.217	2146	.284	1924	.36	1721	.443
5	3560	.144	3050	.196	2650	.256	2375	.32	2125	.4
6	5112	.12	4356	.163	3816	.213	3420	.27	3060	.33
7	6958	.103	5929	.14	5194	.183	4655	.23	4165	.29
8	9088	.09	7744	.123	6784	.16	6080	203	5440	.25
9			9801	.109	8586	.142	7695	.18	6885	.22
10			12100	.098	10600	.128	9500	.162	8500	.2
11					12826	.117	11495	.15	10285	.182
12					15264	.107	13680	.135	12240	.17
13							16100	.125	14400	.154
14							18600	.115	16700	.143

	12 feet.		14 feet.		16 feet.		18 feet.		20 feet.	
6	2548	.48	2184	.65	1912	.85	1699	1.08	1530	1.34
7	3471	.41	2975	.58	2603	.73	2314	.93	2082	1.14
8	4532	.36	3884	.49	3396	.64	3020	.81	2720	1.00
9	5733	.32	4914	.44	4302	.57	3825	.72	3438	.89
10	7083	.28	6071	.39	5312	.51	4722	.64	4250	.8
11	8570	.26	7346	.36	6428	.47	5714	.59	5142	.73
12	10192	.24	8736	.33	7648	.43	6796	.54	6120	.67
13	11971	.22	10260	.31	8978	.39	7980	.49	7182	.61
14	13883	.21	11900	.28	10412	.36	9255	.46	8330	.57
15	15937	.19	13660	.26	11952	.34	10624	.43	9562	.53
16	18129	.18	15536	.24	13584	.32	12080	.40	10880	.5
17	20500	.17	17500	.23	15353	.30	13647	.38	12282	.47
18	22932	.16	19656	.21	17208	.28	15700	.36	13752	.44

NOTE.—This Table shows the greatest weight that ever ought to be laid upon a beam for permanent-load ; and, if there be any liability to jerks, &c., ample allowance must be made ; also, the weight of the beam itself must be included. (*See Tables of Cast Iron.*)

To find the weight of a cast iron beam of given dimensions.

RULE.—Multiply the sectional area in inches by the length in feet, and by 3.2, the product equal the weight in lbs.

EX. Required the weight of a uniform rectangular beam of cast iron, 16 feet in length, 11 inches in breadth, and 1½ inch in thickness.

$$11 \times 1.5 \times 16 \times 3.2 = 844.8 \text{ lbs.}$$

RESISTANCE OF BODIES TO FLEXURE BY VERTICAL PRESSURE.

When a piece of timber is employed as a column or support, its tendency to yielding by compression is different according to the proportion between its length and area of its cross section ; and supposing the form that of a cylinder whose length is less than seven or eight times its diameter, it is impossible to bend it by any force applied longitudinally, as it will be destroyed by splitting before that bending can take place ; but when the length exceeds this, the column will bend under a certain load, and be ultimately destroyed by a similar kind of action to that which has place in the transverse strain. Columns of cast iron and of other bodies are also similarly circumstanced.

When the length of a cast iron column with flat ends equals about thirty times its diameter, fracture will be produced wholly by bending of the material.   When of less length, fracture takes place partly by crushing and partly by bending. But, when the column is enlarged in the middle of its length from one and a half to twice its diameter at the ends, by being cast hollow, the strength is greater by one-seventh than in a solid column containing the same quantity of material.

To determine the dimensions of a support or column to bear, without sensible curvature, a given pressure in the direction of its axis.

Rule.—Multiply the pressure to be supported in lbs. by the square of the column's length in feet, and divide the product by twenty times the tabular value of E ; and the quotient will be equal to the breadth multiplied by the cube of the least thickness, both being expressed in inches.

Note 1.—When the pillar or support is a square, its side will be the fourth root of the quotient.

Note 2.—If the pillar or column be a cylinder, multiply the tabular value of E by 12, and the fourth root of the quotient equal the diameter.

Ex. 1.   What should be the least dimensions of an oak support, to bear a weight of 2240 lbs., without sensible flexure, its breadth being 3 inches, and its length 5 feet?

$$\text{Tabular value of } E = 105,$$
$$\text{and } \frac{2240 \times 5^2}{20 \times 105 \times 3} = \sqrt[3]{8.888} = 2.05 \text{ inches.}$$

Ex. 2   Required the side of a square piece of Riga fir, 9 feet in length, to bear a permanent weight of 6000 lbs.

$$\text{Tabular value of } E = 96,$$
$$\text{and} \frac{6000 \times 9^2}{20 \times 96} = \sqrt[4]{253} = 4 \text{ inches nearly. .}$$

ELASTICITY OF TORSION, OR RESISTANCE OF BODIES TO TWISTING.

The angle of flexure by torsion is as the length and extensibility of the body directly and inversely as the diameter; hence, the length of a bar or shaft being given, the power, and the leverage the power acts with, being known, and also the number of degrees of torsion that will not affect the action of the machine, to determine the diameter in cast iron with a given angle of flexure.

Rule.—Multiply the power in lbs. by the length of the shaft in feet, and by the leverage in feet; divide the product by fifty-five times the number of degrees in the angle of torsion ; and the fourth root of the quotient equal the shaft's diameter in inches.

Ex.   Required the diameters for a series of shafts 35 feet in length, and to transmit a power equal to 1245 lbs., acting at the circumference of a wheel 2½ feet radius, so that the twist of the shafts on the application of the power may not exceed one degree.

$$\frac{1245 \times 35 \times 2.5}{55 \times 1} = \sqrt[4]{1981} = 6.67 \text{ inches in diameter.}$$

To determine the side of a square shaft to resist torsion with a given flexure.

Rule.—Multiply the power in pounds by the leverage it acts with in feet, and also by the length of the shaft in feet; divide this product by 92.5 times the angle of flexure in degrees, and the square root of the quotient equals the area of the shaft in inches.

Ex.   Suppose the length of a shaft to be 12 feet, and to be driven by a power equal to 700 lbs., acting at 1 foot from the centre of the shaft—required the area of cross section, so that it may not exceed 1 degree of flexure.

$$\frac{700 \times 1 \times 12}{92.5 \times 1} = \sqrt[2]{90.8} = 9.53 \text{ inches.}$$

Relative strength of Bodies to resist Torsion, Lead being 1.

Tin 1.4	Gun Metal.........5.0	English Iron........10.1
Copper 4.3	Cast Iron 9.0	Blistered Steel 16 6
Yellow Brass 4.6	Swedish Iron 9.5	Shear Steel 17.0

STRENGTH OF MATERIALS.

[From Grier's Mechanic's Calculator, &c.]

BAR OF IRON.—The average breaking weight of a Bar of Wrought Iron, 1 inch square, is 25 tons ; its elasticity is destroyed, however, by about two-fifths of that weight, or 10 tons. It is extended, within the limits of its elasticity, .000096, or one-tenthousandth part of an inch for every ton of strain per square inch of sectional area. Hence, the greatest constant load should never exceed one-fifth of its breaking weight, or 5 tons for every square inch of sectional area.

The lateral strength of wrought iron, as compared with cast iron, is as 14 to 9. Mr. Barlow finds that wrought iron bars, 3 inches deep, 1 1-2 inches thick, and 33 inches between the supports, will carry 4 1-2 tons.

BRIDGES.—The greatest extraneous load on a square foot is about 120 pounds.

FLOORS.—The least load on a square foot is about 160 pounds.

ROOFS.—Covered with slate, on a square foot, 51 1-2 pounds.

BEAMS.— When a beam is supported in the middle and loaded at each end, it will bear the same weight as when supported at both ends and loaded in the middle ; that is, each end will bear half the weight.

Cast Iron Beams should not be loaded to more than one-fifth of their ultimate strength.

The strength of similar *beams* varies inversely as their lengths ; that is, if a beam 10 feet long will support 1000 pounds, a similar beam 20 feet long would support only 500 pounds.

A *beam* supported at one end will sustain only one-fourth part the weight which it would if supported at both ends.

When a *beam* is fixed at both ends, and loaded in the middle, it will bear one-half more than it will when loose at both ends. When the beam is loaded uniformly throughout it will bear double. When the beam is fixed at both ends, and loaded uniformly, it will bear triple the weight.

In any *beam* standing obliquely, or in a sloping direction, its strength or strain will be equal to that of a beam of the same breadth, thickness, and material, but only of the length of the horizontal distance between the points of support.

In the construction of *beams*, it is necessary that their form should be such that they will be equally strong throughout. If a beam be fixed at one end, and loaded at the other, and the breadth uniform throughout its length, then, that the beam may be equally strong throughout, its form must be that of a parabola. This form is generally used in the beams of steam engines.

When a *beam* is regularly diminished towards the points that are least strained, so that all the sections are similar figures, whether it be supported at each end and loaded in the middle, or supported in the middle and loaded at each end, the outline should be a cubic parabola.

When a beam is supported at both ends, and is of the same breadth throughout, then, if the load be uniformly distributed throughout the length of the beam, the line bounding the compressed side should be a semi-ellipse.

The same form should be made use of for the rails of a wagon-way, where they have to resist the pressure of a load rolling over them.

Similar *plates* of the same thickness, either supported at the ends or all round, will carry the same weight either uniformly distributed or laid on similar points, whatever be their extent.

13

The lateral strength of any *beam*, or *bar* of *wood, stone, metal*, &c., is in proportion to its breadth multiplied by its depth[2]. In square beams the lateral strengths are in proportion to the cubes of the sides, and in general of like-sided beams as the cubes of the similar sides of the section.

The lateral strength of any *beam* or *bar*, one end being fixed in the wall and the other projecting, is inversely as the distance of the weight from the section acted upon; and the strain upon any section is directly as the distance of the weight from that section.

The absolute strength of *ropes* or *bars*, pulled lengthwise, is in proportion to the squares of their diameters. All cylindrical or prismatic rods are equally strong in every part, if they are equally thick, but if not they will break where the thickness is least.

The strength of a *tube*, or *hollow cylinder*, is to the strength of a solid one as the difference between the fourth powers of the exterior and interior diameters of the tube, divided by the exterior diameter, is to the cube of the diameter of a solid cylinder,— the quantity of matter in each being the same. Hence, from this it will be found, that a hollow cylinder is one-half stronger than a solid one having the same weight of material.

The strength of a column to resist being crushed is directly as the square of the diameter, provided it is not so long as to have a chance of bending. This is true in metals or stone, but in timber the proportion is rather greater than the square.

MODELS PROPORTIONED TO MACHINES.

The relation of models to machines, as to strength, deserves the particular attention of the mechanic. A model may be perfectly proportioned in all its parts as a model, yet the machine, if constructed in the same proportion, will not be sufficiently strong in every part; hence, particular attention should be paid to the kind of strain the different parts are exposed to; and from the statements which follow, the proper dimensions of the structure may be determined.

If the strain to draw asunder in the model be 1, and if the structure is 8 times larger than the model, then the stress in the structure will be 8^3 equal 512. If the structure is 6 times as large as the model, then the stress on the structure will be 6^3 equal 216, and so on; therefore, the structure will be much less firm than the model; and this the more, as the structure is cube times greater than the model. If we wish to determine the greatest size we can make a machine of which we have a model, we have,

The greatest weight which the beam of the model can bear, divided by the weight which it actually sustains equal a quotient which, when multiplied by the size of the beam in the model, will give the greatest possible size of the same beam in the structure.

Ex.—If a beam in the model be 7 inches long, and bear a weight of 4 lbs. but is capable of bearing a weight of 26 lbs.; what is the greatest length which we can make the corresponding beam in the structure? Here

$$26 \div 4 = 6.5, \qquad \text{therefore, } 6.5 \times 7 = 45.5 \text{ inches.}$$

The strength to resist crushing, increases from a model to a structure in proportion to their size, but, as above, the strain increases as the cubes; wherefore, in this case, also, the model will be stronger than the machine. and the greatest size of the structure will be found by employing the square root of the quotient in the last rule, instead of the quotient itself; thus,

If the greatest weight which the column in a model can bear is 3 cwt., and if it actually bears 28 lbs., then, if the column be 18 inches high, we have

$$\sqrt{\left(\frac{336}{28}\right)} = 3.464 ; \qquad \text{wherefore } 3.464 \times 18 = 62.352$$

inches, the length of the column in the structure.

STRENGTH OF MATERIALS.

[*From Adcock's Engineer.*]

List of metals, arranged according to their strength.—Steel, wrought-iron, cast-iron, platinum, silver, copper, brass, gold, tin, bismuth, zinc, antimony, lead.

According to Tredgold's and Duleau's experiments, a piece of the best bar-iron 1 square inch across the end would bear a weight of about 77,373 lbs., while a similar piece of cast-iron would be torn asunder by a weight of from 16,243 to 19,464 lbs. Thin iron wires, arranged parallel to each other, and presenting a surface at their extremity of 1 square inch, will carry a mean weight of 126,340 lbs.

List of woods, arranged according to their strength.—Oak, alder, lime, box, pine (*sylr.*), ash, elm, yellow pine, fir.

A piece of well-dried pine wood, presenting a section of 1 square inch, is able, according to Eytelwein, to support a weight of from 15,646 lbs. to 20,400 lbs., whilst a similar piece of oak will carry as much as 25,850 lbs.

Hempen cords, twisted, will support the following weights to the *square inch* of their section :

¼-inch to 1 inch thick, 8,746 lbs.; 1 to 3 inches thick, 6,800 lbs.; 3 to 5 inches thick, 5,345 lbs.; 5 to 7 inches thick, 4,860 lbs.

Tredgold gives the following rule for finding the weight in lbs. which a hempen rope will be capable of supporting : Multiply the square of the circumference in inches by 200, and the product will be the quantity sought.

In the practical application of these measures of absolute strength, that of metals should be reckoned at one-half, and that of woods and cords at one-third of their estimated value.

In a parallelopipedon of uniform thickness, supported on two points and loaded in the middle, *the lateral strength is directly as the product of the breadth into the square of the depth, and inversely as the length.* Let W represent the lateral strength of any material, estimated by the weight, *b* the breadth, and *d* the depth of its end, and *l* the distance between the points of support; then $W = f d^2 b \div l$.

If the parallelopipedon be fastened only at one end in a horizontal position, and the load be applied at the opposite end, $W = f d^2 b \div 4l$.

It is to be observed that the three dimensions, *b*, *d*, and *l*, are to be taken in the same measure, and that *b* be so great that no lateral curvature arise from the weight ; *f* in each formula represents the lateral strength, which varies in different materials, and which must be learnt experimentally.

A beam having a rectangular end, whose breadth is two or three times greater than the breadth of another beam, has a power of suspension respectively two or three times greater than it ; if the end be two or three times deeper than the end of the other, the suspension power of that which has the greater depth exceeds the suspension power of the other, four or nine times ; if its length be two or three times greater than the length of another beam, its power of suspension will be ⅓ or 1·3 respectively that of the other.; provided that in each case the mode of suspension. the position of the weight, and other circumstances be similar. Hence it follows that a beam, one of whose sides tapers, has a greater power of suspension if placed on the slant than on the broad side, and that the powers of suspension in both cases are in the ratio of their sides ; so, for instance, a beam, one of whose sides is double the width of the other, will carry twice as much if placed on the narrow side, as it would if laid on the wide one.

In a piece of round timber (a cylinder) the power of suspension is in proportion to the diameters cubed, and inversely as the length ; thus a beam with a diameter two or three times longer than that of another, will carry a weight 8 or 27 times heavier respectively than that whose diameter is unity, the mode of fastening and loading it being similar in both cases.

The lateral strength of square timber is to that of a tree whence it is hewn as 10 : 17 nearly.

A considerable advantage is frequently secured by using hollow cylinders instead of solid ones, which, with an equal expenditure of materials, have far greater strength, provided only that the solid part of the cylinder be of a sufficient thickness, and that the workmanship be good ; especially that in cast metal beams be the thickness be uniform, and the metal free from flaws. According to Eytelwein, such hollow cylinders are to solid ones of equal weight of metal as 1.212 : 1, when the inner semi-diameter is to the outer as 1 : 2 ; according to Tredgold as 17 : 10, when the two semi-diameters are to each other as 15 : 25, and as 2 : 1, when they are to each other as 7 : 10.

A method of increasing the suspensive power of timber supported at both ends, is, to saw down from $\frac{1}{8}$ to $\frac{1}{2}$ of its depth, and forcibly drive in a wedge of metal or hard wood, until the timber is slightly raised at the middle out of the horizontal line. By experiment it was found that the suspensive power of a beam thus cut 1-3 of its depth was increased 1-19th, when cut $\frac{1}{2}$ it was increased 1-29th, and when cut 3-4th through it was increased 1-87th.

The force required to crush a body increases as the section of the body increases ; and this quantity being constant, the resistance of the body diminishes as the height increases.

According to Eytelwein's experiments, the strength of columns or timbers of rectangular form in resisting compression is, as

1. The cube of their thickness (the lesser dimension of their section). 2. As the breadth (the greater dimension of their section). 3. inversely as the square of their length.

Cohesive power of Bars of Metal one inch square, in Tons.

Iron, Swedish bar.	29.20	Copper, wrought . . .	15.08
Do., Russian bar	26.70	Gun metal	16.23
Do., English bar	25.00	Copper, cast	8.51
Steel, cast	59.93	Brass, cast, yellow . . .	8.01
Do., blistered	59.43	Iron, cast	7.87
Do., sheer	56.97	Tin, cast	2.11

RELATIVE STRENGTH OF CAST AND MALLEABLE IRON.

It has been found, in the course of the experiments made by Mr. Hodgkinson and Mr. Fairbairn, that the average strain that cast iron will bear in the way of tension, before breaking, is about seven tons and a half per square inch ; the weakest, in the course of 16 trials on various descriptions, bearing 6 tons, and the strongest 9 3-4 tons. The experiments of Telford and Brown show that malleable iron will bear, on an average, 27 tons ; the weakest bearing 24, and the strongest 29 tons. On approaching the breaking point, cast iron may snap in an instant, without any previous symptom, while wrought iron begins to stretch, with half its breaking weight, and so continues to stretch till it breaks. The experiments of Hodgkinson and Fairbairn show also that cast iron is capable of sustaining compression to the extent of nearly 50 tons on the square inch ; the weakest bearing 36½ tons, and the strongest 60 tons. In this respect, malleable iron is much inferior to cast iron. With 12 tons on the square inch it yields, contracts in length, and expands laterally ; though it will bear 27 tons, or more, without actual fracture.

Rennie states that cast iron may be crushed with a weight of 93,000 lbs., and brick with one of 562 lbs. on the square inch.

STRENGTH OF BEAMS.

[From Lowndes' Engineer's Hand-book,—Liverpool, 1860.]

SOLID, RECTANGULAR, AND ROUND : TO FIND THEIR STRENGTH.

Square and rectangular.

$$\frac{(\text{Depth ins.})^2 \times \text{Thickness ins.}}{\text{Length, ft.}} \times \text{Tabular No.} = \text{Breaking weight, tons.}$$

Round.

$$\frac{(\text{Diameter ins.})^3}{\text{Length in ft.}} \times \text{Tabular No.} = \text{Breaking weight, tons.}$$

Hollow.

$$\frac{(\text{Outside dia. ins.})^3 - (\text{Inside dia. ins.})}{\text{Length, ft.}} \times \text{Tabular No.} = \text{Breaking weight}$$

tons.

Thickness not exceeding {	1 inch for iron. 3 ins. for wood.	2 ins. for iron. 6 ins. for wood.	3 ins. for iron. 12 ins. for wood.

Square and Rectangular.

Cast and Wrought Iron	1	·85	·7
Teak and greenheart	·36	·32	·26
Pitch pine, and Canadian oak	·25	·22	·18
Fir, red pine, and English oak	·18	·16	·13

Round.

Cast and Wrought Iron	·8	·68	·56
Teak and greenheart .	·28	·25	·2
Fir and English oak .	·14	·125	·1

To find the Breaking Weight in lbs. use the Tabular No. below.

Thickness not exceeding {	1 inch for iron. 3 ins. for wood.	2 ins. for iron. 6 ins. for wood.	3 ins. for iron. 12 ins. for wood.

Square and Rectangular.

Iron	2240	1900	1570
Teak	800	710	570
Fir and oak	400	355	285

13*

Round.

Iron	1800	1570	1260
Teak	640	570	460
Fir and oak	320	285	230

Though wrought and cast iron are represented in these rules as of equal strength, it should be observed that while a cast iron bar 1 inch × 1 inch × 1 foot 0 inch long, of average quality, will break with one ton, a similar bar of wrought iron only loses its elasticity, and deflects 1-16th of an inch, yet as it can only carry a further weight by destroying its shape and increasing the deflection, it is best to calculate on the above basis :—

A wrought iron bar 1 in. × 1 in. × 1 ft. 0 in. long } deflects
$$\begin{array}{l} 1\text{-}16 \text{ with } 1 \text{ ton.} \\ 1\text{-}8 \quad\text{``}\quad 1\tfrac{1}{2} \quad\text{``} \\ 2\;1\text{-}2 \quad\text{``}\quad 2\tfrac{1}{4} \quad\text{``} \end{array}$$

The above rule gives the weight that will break the beam if put on the middle. If the weight is laid equally all over, it would require double the weight to break it.

A beam should not be loaded with more than 1-3 of the breaking weight in any case, and as a general rule not with more than 1-4, for purposes of machinery not with more than 1-6 to 1-10 depending on circumstances.

To find the proper size for any given purpose.

Rectangular.

$$\frac{\text{Weight} \times \text{Length ft.}}{\text{Tabular No.}} \times 3 \text{ or } 4 \text{ or } 6, \&\text{c. according to circumstances} = \text{B } \text{D}^2 \text{ ins.}$$

Round.

$$\sqrt[3]{\frac{\text{Weight} \times \text{Length. ft.}}{\text{Tabular No.}}} \times 3 \text{ or } 4 \text{ or } 6, \&\text{c. according to circumstances} = \text{diam. ins.}$$

CAST IRON WITH FEATHERS OR FLANGES : TO FIND THEIR STRENGTH.

$$\frac{\text{Sec: area, bottom flange ins.} \times \text{depth ins.}}{\text{Length in feet.}} \times 2 = \text{Breaking weight, tons.}$$

If the metal exceeds 1 inch in thickness deduct 1-8th.

If above 2 inches deduct 1-4th.

This description of beam is of the strongest form, when the sectional area of the bottom flange is six times that of the top flange.

In designing this description of beam, the bottom flange may be from 1-2 to 1 1-2 the depth of beam; the top flange from 1-4 to 1-3 the width of the bottom one, and 2-3 to 1-2 the thickness of it; the feather being made at the top a little thicker than the top flange, increasing to the bottom to nearly the thickness of the bottom flange; in this way avoiding any sudden variation in the thickness and saving weight; many engineers, however, prefer keeping the same thickness throughout in every part. The vertical brackets for stiffening the girder should not be made straight, but hollowed out something like the sketch, as thus they are much less liable to crack, and all the corners should be well filled in.

In most cases it is necessary that the beam should be of uniform depth·throughout; it will, however, save weight, without diminishing the strength of the beam, if the width of the bottom flange be reduced very considerably towards the ends; 1-2 of the width of the middle being quite sufficient; care being taken to maintain a sufficient surface for bearing, if the beam has to be carried on a wall,

Fig. 1.

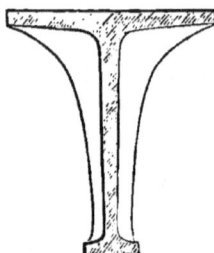

Fig. 2.

WROUGHT IRON BEAMS.

Girders.—The sketch shows a very strong form for this description of girder, when rolled solid.   The top flange being condensed and square is in a good form to resist compression; the bottom flange has a wider surface to rest on, and the middle rib is light ;

an experimental beam of this description 8 ins. deep and 11 feet long requiring 5 tons to break it.

The top flange should have a sectional area 1 1-2 times that of the bottom. When thus proportioned :

$$\frac{\text{Sec. area top flange, ins.} \times \text{depth ins.}}{\text{Length feet.}} \times 5 = \text{Breaking weight in tons.}$$

This is an inferior shape.   *Fig.* 4.

In such a beam the top flange should have an area 1 3-4 that of the bottom flange.

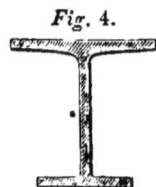

When thus proportioned :

$$\frac{\text{Sec. area top flange ins.} \times \text{depth ins.}}{\text{Length feet.}} \times 4 = \text{Breaking}$$

weight, tons.

Beams of the above forms, made of plates and of L iron, are of equal strength with the above; care being taken to make the bottom flange of double plates, with joint plates over the butts, allowing a little extra area in the bottom to compensate for the rivet holes, though this is not necessary if they are rivetted up by steam.

WROUGHT IRON BEAMS.

Fig. 5.

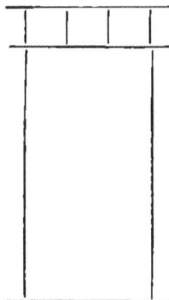

Hollow Girders.—The sketch represents the form for hollow girders combining the greatest strength with the least weight, the top being in the best form for resisting compression.

The proportion of the bottom sectional area to that of the top should be as 11 to 12, or 4-5 ; and the sides should be well stiffened with angle iron, to keep them from buckling ; the sectional area of the top and bottom may be reduced at the extremities to 1-3 of the area at the middle, without diminishing the strength of the beam.

When thus proportioned :

$$\frac{\text{Section. area top, ins.} \times \text{depth ins.}}{\text{Length feet.}} \times 5 = \text{Breaking weight, tons.}$$

An experimental beam of this form, 75 feet long between supports, 4 feet 6 inches deep, with 6 cells at the top, about 6 inches square each, with a sectional area 24 sq. ins., the sides stiffened with 1 1-2 L irons, 2 feet apart, required 86 tons to break it.

Fig. 6.

In the plain hollow girder the top should have a sectional area 1 3-4 that of the bottom.

Thus proportioned :

$$\frac{\text{Section. area top, ins.} \times \text{depth ins.}}{\text{Length feet.}} \times 4 = \text{Breaking weight tons.}$$

To find the strength of a round girder.

$$\frac{\text{Sec. area, ins.} \times \text{dia. ins.}}{\text{Length feet.}} = \text{Breaking weight, tons.}$$

To find the strength of any beam.

If the top flange is the weakest, find the compressive breaking strain in tons per square inch due to its shape, thickness, and length. (See Columns.)

If the bottom is the weakest, find the tensional breaking strain of the material in tons per square inch.

Then,

$$\frac{\text{Sec. area ins. of weakest flange} \times \text{breaking strain, tons per in.} \times \text{depth of beam ft.} \times 4}{\text{Length between supports, feet.}}$$

= Breaking weight, tons.

This rule will be found useful, either to confirm the results obtained from the previous rules, or to find the strength of any beams of irregular shape not included in them.

The mode of ascertaining the compression and tension on the top and bottom flanges of beams is sufficiently simple.

Take the case of a beam, 20 feet long, 2 feet deep, with a weight of 20 tons on the middle ; the force counteracting this weight will be 10 tons on each end; the force of compression at the top in the middle of the beam, and that of tension at the bottom, taking the central weight as the fulcrum, will be just in proportion to the leverage; in this case, as 10 to 2, or 5 to 1. The force of 10 tons applied to the end will thus result in a force of 50 tons of compression and tension on the flanges in the middle of the beam. Or in a simple form,

$$\frac{\text{Weight, tons} \times \text{length, feet}}{\text{Depth, feet} \times 4} = \text{Strain on top and bottom flanges, tons.}$$

The ultimate compressive strength of boiler plate iron may be taken at 16 tons per square inch, the tensile strength at 20 tons per square inch; and this is the reason why, in all wrought iron beams, the top requires to be the strongest.

But as in cast iron the compressive strength is about 48 tons, while the tensile strength is only about 7 tons per square inch, the bottom flange in cast iron girders requires to be much the strongest.

The fullest information on this subject, and the experiments in detail, will be found in Mr. Eaton Hodgkinson's experiments on the strength of cast iron beams, and in Mr. Edwin Clark's work on the Britannia and Conway tubular bridges.

SOLID COLUMNS.

Fail by crushing with length under - - - - - - - - - 5 diameters.
Principally by crushing from - - - - - - - - - 5 to 15 "
Partly by crushing, partly by bending, from - - - 15 to 25 "
Altogether by bending above - - - - - - - - - 25 "

Cast iron of average quality is crushed with - - 49 tons per square inch.
Wrought iron of average quality is crushed with 16 " " "
Wrought iron is permanently injured with - - - 12 " " "
Oak wrought is crushed with - - - - - - - 4 " " "
Deal wrought is crushed with - - - - - - - 2 " " "

The comparative strength of different columns, of different lengths, will be seen very clearly from the following table derived from experiments by Mr. Hodgkinson :—

Wrought Iron Bars.		Proportion of Length to Thickness.	Gave way with
Square.	Length.		
ins.	ft. ins.		
1 × 1	7½	7½ to 1	21·7 tons per sq. inch
"	1 3	15 to 1	15·4 "
"	2 6	30 to 1	11·3 "
"	5 0	60 to 1	7·5 "
"	7 6	90 to 1	4·3 "
½ × ½	5 0	120 to 1	2·5 "
"	7 6	180 to 1	1· "

To find the strength of any wrought iron column with square ends.

Area of column sq. inches × tons per inch corresponding to proportion of length, as per table above = Breaking weight, tons.

If the ends are rounded, divide the final result by 3 to find the breaking weight.

In columns of oblong section, the narrowest side must always be taken in calculating the proportion of height to width.

To find the strength of round columns exceeding 25 diameters in length.
Mr. Hodgkinson's rule.

$$\frac{(\text{Diameter, ins.})^{3\cdot6}}{(\text{Length, ft.})^{1\cdot7}} \times \text{Tabular No.} = \text{Breaking weight, tons.}$$

	Square Ends.	Rounded or Moveable Ends.
Wrought iron - - - - -	77	26
Cast iron - - - - - -	44	15
Dantzic oak - - - - -	4·5	1 7
Red deal - - - - - -	3·3	1·2

A column should not be loaded with more than 1-3 of the breaking weight in any case, and as a general rule, not with more than 1-4 ; for purposes of machinery not with more than 1-6 to 1-10, according to circumstances.

Tables of Powers for the Diameters and Lengths of Columns.

Diameter.	3·6 Power.	Diameter.	3·6 Power.		Length.	1·7 Power.
1 in.	1·	7 in.	1102·4		1	1·
¼	2·23	¼	1251·		2	3·25
½	4·3	½	1413·3		3	6·47
¾	7·5	¾	1590·3		4	10·556
2	12·1	8	1782·9		5	15·426
¼	18·5	¼	1991·7		6	21·031
½	27·	½	2217·7		7	27·332
¾	38·16	¾	2461·7		8	34·297
3	52·2	9	2724·4		9	41·9
¼	69·63	¼	3006·85		10	50·119
½	90·9	½	3309·8		11	58·934
¾	116·55	¾	3634·3		12	68·329
4	147·	10	3981·07		13	78·289
¼	182·9	¼	4351·2		14	88·8
½	224·68	½	4745·5		15	99·85
¾	272·96	¾	5165·		16	111·43
5	328·3	11	5610 7		17	123·53
¼	391·36	¼	6083·4		18	136·13
½	462·71	· ½	6584·3		19	149·24
¾	543·01	¾	7114·4		20	162·84
6	632·91	12	7674·5		21	176·92
¼	733·11				22	191·48
½	844·28				23	206·51
¾	967·15				24	222·

HOLLOW COLUMNS.

Hollow columns fail principally by crushing, provided the length does not exceed 25 diameters; indeed, the length does not appear to affect the strength much till it exceeds 50 diameters.

The comparative strength of different forms and of different thicknesses will appear so distinctly from the experiments below, made by Mr. Hodgkinson, that no difficulty will be found in ascertaining the strength due to any size or form of column that may be required.

Square Columns of Plate Iron Rivetted

Columns 10 ft. 0 in. long.				
Size.	Thickness.	Proportion of Thickness to Width.	Proportion of Length to Width.	Break'g weight Tons per sq. in. of section.
4 in. $\times$ 4 in.	·03	$\frac{1}{133}$	30 to 1	4·9
"	·06	$\frac{1}{66}$	"	8·6
"	·1	$\frac{1}{40}$	"	10·
"	·2	$\frac{1}{20}$	"	12·
8 in. $\times$ 8 in.	·06	$\frac{1}{133}$	15 to 1	6·
"	·14	$\frac{1}{66}$	"	9·
"	·22	$\frac{1}{36}$	"	11·5
"	·25	$\frac{1}{32}$	"	12·
Column 8 feet 0 inches long.				
18 $\times$ 18	·5	$\frac{1}{30}$ practically	5·4 to 1	13·6
Column 10 feet 0 inches long, with Cells.				
8 in. $\times$ 8 in.	·06	$\frac{1}{66}$ of width of cells	15 to 1	8·6

To find the strength of any Hollow Wrought Iron Column.

Sec. area, sq. ins. $\times$ Tons per inch, corresponding to the proportions of length and thickness to width as per tables $=$ Breaking weight, tons.

Columns of Oblong Section.

The strength of these may be ascertained by the same rule as that of square columns. The smallest width being taken in calculating the proportion of height to width, while the longest side must be taken into consideration in calculating the proportion of thickness to width.

Column 10 feet 0 inches long.

Size.	Thickness.	Proportion of Thickness to greatest Width.	Proportion of Length to least Width.	Actual Breaking Weight Tons per sq. in. of Section.
8 in. $\times$ 4 in.	·06	$\frac{1}{133}$	30 to 1	6·78

Round. Columns of Plate Iron Rivetted.

					Same Columns Reduced in Length.		
	Columns 10 ft. 0 in. long.						
Diameter.	Thickness.	Proportion of thickness to Diameter.	Proportion of length to Diameter.	Breaking Weight. Tons per sq. inch.	Breaking Weights. Tons per square inch.		
					5 ft. 0 in. long.	2 ft. 6 in. long.	
1½	·1	$\frac{1}{15}$	80 to 1	6·5	13·9	5·8
2	·1	$\frac{1}{20}$	60 to 1	10·35	14·8	16·5
2½	·1	$\frac{1}{25}$	48 to 1	13·3	15·6	16·3
2½	·24	$\frac{1}{11}$	48 to 1	9·6	15·6	16·
2½	·21	$\frac{1}{12}$	48 to 1	9·9	13·	17·
3	·15	$\frac{1}{20}$	40 to 1	12·36	13·	16·5
4	·15	$\frac{1}{26}$	30 to 1	12·34	13·	
6	·1	$\frac{1}{60}$	20 to 1	15·	17·	18·6
6	·13	$\frac{1}{46}$	20 to 1	18·6		

It would seem from this that a thickness of 1-48, or 1-4 inch in thickness for every foot in diameter is a good proportion for this kind of column.

It will be seen from these experiments, that it is the proportion of thickness to the width of cell which regulates the strength within certain limits of height.

And that a thickness of 1-30 or 1-8 inch for every 4 inches in width will give the highest result practicable for square columns.

CRANE.

The strains on the principal parts can be ascertained with great ease in the following manner—the strength being proportioned accordingly.

To find the strain on the post.

$$\frac{\text{Weight suspended, tons} \times \text{Projection, feet}}{\text{Height of post above ground, feet}} = \text{Strain on top of post, tons.}$$

The post can then be calculated as a beam, twice as long as this height from ground, with twice the weight on the middle. [*See Beams.*]

COLD WATER PUMP.

Usually 1-4 of cylinder diameter when the stroke is 1-2 that of piston.
    1-3   "   "   1-4   "

To find the proper size, under any circumstances, capable of supplying twice the quantity ordinarily used for injection.

$$\frac{\text{Cub. ft. water per hour used in cylinder in form of steam}}{\text{Stroke of pump, ft.} \times \text{strokes per minute}} = \text{Area of pump}$$
in square feet.

FAN.

Case should be strong and heavy.   Bearings long.
Blades and arms as light and well balanced as possible.
Good proportions —

> Inlet $= \frac{1}{2}$ diameter of fan,
> Blades $= \frac{1}{4}$ diameter of fan each way,
> Outlet $=$ area of blades.

The area of tuyeres is most advantageous when made

$$= \frac{\text{area of blades}}{\text{density of blast, oz. per sq. inch.}}$$

and it should not exceed double this size.

The best Velocity of Circumference for different Densities.

Velocity of Circumference. Feet per Second.	Density of Blast. Oz. per inch.
170	3
180	4
195	5
205	6
215	7

A speed of 180 to 200 feet per second, giving a density of 4 or 5 oz., is very suitable for smithy fires.

250 to 300 feet per second is a proper speed for cupolas.

A fan 4 feet 0 inch diameter, blade 1 foot 0 inch square, will supply 40 fires with $1\frac{5}{8}$ tuyeres at a density of 4 oz.

To find the Horse Power required for any fan,

Let D $=$ density of blast in oz. per inch.
> A $=$ area of discharge at tuyeres in square inches.
> V $=$ velocity of circumference in feet per second.

Then $\dfrac{\dfrac{V^2}{1000} \times D \times A}{963} =$ Effective Horse Power required.

To find the density to be attained with any given fan.

Let D $=$ diameter of fan in feet.

Then $\dfrac{\left(\dfrac{V}{4}\right)^2}{120 \times d.} =$ Density of blast in oz. per inch.

Or the density may be found by comparison with the following table :—

14

Velocity of Circumference. Feet per Second.	Area of Nozzles.		Density of Blast. Oz. per inch.
150	Twice area of blades		1
150	Equal	ditto	2
150	1-2	ditto	3
170	1-4	ditto	4
200	1-2	ditto	4
200	1-6	ditto	6
220	1-3	ditto	6

To find the quantity of air that will be delivered by any Fan, the density being known.

Total area nozzles, sq. ft. $\times$ velocity, ft. per minute corresponding to density (as per table) = Air delivered, cubic ft. per minute.

Density. Oz. per Sq. Inch.	Velocity. Feet per Minute.	Density. Lbs. per Sq. Inch	Velocity. Feet per Minute.
1	5,000	1	20,000
2	7,000	$1\frac{1}{2}$	24,500
3	8,600	2	28,300
4	10,000	$2\frac{1}{2}$	31,600
5	11,000	3	44,640
6	12,250	4	40,000
7	13,200	6	49,000
8	14,150	8	56,600
9	15,000	10	63,200
10	15,800	12	69,280
11	16,500	15	78,000
12	17,300	20	89,400

FRICTION.

From Mr. Rennie's Experiments.

The friction of metal on metal, without unguents,
May be taken at 1-6 of the weight up to 40 lbs. per sq. in.
     "   1-5   "   "   100   "
Brass on cast iron 1-4   "   "   800   "
Wrought on cast iron 1-3 "   "   500   "
    With tallow at   1-10 of the weight.
    " olive oil at   1-13   "
800 lbs. per inch forces out the oil.
Friction of journals under ordinary circumstances 1-30 of weight.
    "   well oiled, sometimes only   1-60   "

CENTRIFUGAL FORCE.

$$\frac{(\text{Revolutions per min.})^2 \times \text{dia. in ft.} \times \text{weight}}{5870} = \text{Centrifugal force}$$

in terms of weight.

PEDESTAL — BRACKET.

PEDESTAL.

Good proportions.

Thickness of cover   ·4   of diameter of bearing.
"   of sole plate   ·3   "   "
Diameter of bolts   ·25   "   "   if 2.
"   "   ·18   "   "   if there are 4.

Distance between bolts twice diameter of bearing.

BRACKET.

Solid.   Metal round brass equal to 1-2 diameter of bearing.
General thickness web, &c. equal to 1-4 diameter of bearing.
With feathers.   Width at lightest equal to diameter of bearing.
Thickness equal to 1-6   "   "

TEMPERING.

The article after being completed, is hardened by being heated gradually to a bright red, and then plunged into cold water; it is then tempered by being warmed gradually and equably, either over a fire, or on a piece of heated metal till of the color corresponding to the purpose for which it is required, as per table below, when it is again plunged into water.

Corresponding Temperature.

A very pale straw -   430°   Lancets }
Straw   -   -   -   450°   Razors }
Darker straw -   -   470°   Penknives } All kinds of wood tools
Yellow -   -   -   490°   Scissors }   Screw taps.
Brown yellow   -   500°) Hatchets, Chipping Chisels,
Slightly tinged purple 520° {   Saws.
Purple -   -   -   530°) All kinds of percussive tools.
Dark purple -   -   550° }
Blue   -   •   -   570° } Springs.
Dark blue   -   -   600°) Soft for saws.

To Temper by the Thermometer.

Put the articles to be tempered into a vessel containing a sufficient quantity to cover them, of Oil or Tallow; Sand; or a mixture of 8 parts bismuth, 5 of lead, and 3 of tin, the whole to be brought up to, and kept up at the heat corresponding to the hardness required, by means of a suitable thermometer, till heated equally throughout; the articles are then withdrawn and plunged into cold water.

If no thermometer is available, it may be observed that oil or tallow begins to smoke at 430° or straw color, and that it takes fire on a light being presented, and goes out when the light is withdrawn, at 570° or blue.

CASE HARDENING.

Put the articles requiring to be hardened, after being finished but not polished, into an iron box in layers with animal carbon, that is, horns, hoofs, skins, or leather, partly burned so as to be capable of being reduced to powder, taking care that every part of the iron is completely surrounded; make the box tight with a lute of sand and clay in equal parts, put the whole into the fire, and keep it at a light red heat for half an hour to two hours, according to the depth of hardened surface required, then empty the contents of the box into water, care being taken that any articles liable to buckle be put in separately and carefully, end in first.

Cast iron may be case hardened as follows :—

Bring to a red heat, and roll it in a mixture of powdered prussiate of potash, saltpetre and sal-ammoniac in equal parts, then plunge it into a bath containing 2 oz. prussiate of potash, and 4 oz. sal-ammoniac per gallon of water.

HEAT.

EFFECTS OF HEAT AT CERTAIN TEMPERATURES.—GRIER.

Tin and Bismuth, equal parts, melt at 283 degrees, Fahrenheit; tin melts at 442; polished steel acquires straw color at 460; bismuth melts at 476; sulphur burns at 560; oil of turpentine boils at 560; polished steel acquires deep blue color at 580; lead melts at 594; linseed oil boils at 600; quicksilver boils at 660; zinc melts at 700; iron, bright red in the dark at 752; iron, red-hot in twilight at 884; red heat fully visible in daylight at 1077; brass melts at 3807; copper melts at 4587; silver melts at 4717; gold melts at 5237; welding heat of iron, from 12777; welding heat of iron, to 13427; greatest heat of smith's forge 17327; cast iron begins to melt at 17977; cast iron thoroughly melted at 20577.

SOLDERING.

The solder for joints requires to be of some metal more fusible than that of the substances to be joined.

For Copper, usual solder 6 to 8 parts brass to 1 of zinc; 1 of tin sometimes added.

A still stronger solder, 3 parts brass, 1 of zinc.

To prepare this solder.—Melt the brass in a crucible, when melted add in the zinc, and cover over for 2 or 3 minutes till the combination is effected, then pour it out, over a bundle of twigs, into a vessel of water, or into a mould composed of a number of little channels, so that the solder may be in long strips convenient for use.

Brass filings alone will answer very well.

To braze with this solder.—Scrape the surfaces perfectly clean, and secure the flange or joint carefully; cover the surfaces to be brazed with borax powder moistened; apply the solder, and melt it in with the flame of a clear coke fire from a smith's hearth; particular care being taken not to burn the copper.

Iron and brass are soldered with spelter, which is brass and zinc in equal parts; the process being performed in a manner similar to the above. For ironwork, however, sometimes rather differently; the articles are fixed in their position, and the solder applied, a covering of loam is then put over all to exclude the air, the work thus prepared is then put into the fire a sufficient time to melt the solder in.

BORING AND TURNING.

The best speed for boring cast iron is about $7\frac{3}{4}$ feet per minute.

For drilling about 10 or 11 feet per minute is a good speed for the circumference of the tool. For a 1 inch drill 40 revolutions = 11 feet per minute, other sizes in proportion.

For turning, the proper speed for the circumference is about 15 feet per minute.

BRASS.

COMPOSITIONS OF BRASS.

	Copper.	Tin.	Zinc.
Watch-makers brass	1 part	—	2 parts
German brass	1 "	—	1 "
Yellow brass	2 "	—	1 "
Speculum metal	2 "	1 part	—
Bell metal	3 "	1 "	—
Light castings and small bearings . . .	4 "	1 "	$\frac{1}{4}$ "
Ditto a little harder	4 "	1 "	$\frac{1}{2}$ "
Heavy castings	6 to 7	1 "	1 "
Gun metal	9 "	1 "	—

The addition of a little lead makes the metal more easily wrought, and is advantageous when the work is not intended for exposure to heat.

BRASS CASTING.

As it is often useful to engineers, especially abroad, to be able to cast brass, a slight description of the process may not be out of place.

The ordinary furnace used is of very simple construction.

After lighting the fire, put the pot intended for use bottom upwards over it, so as to warm gradually through. As soon as the fire is burned well through, put the pot into its place, resting the bottom on a fire brick to keep it off the bars, and filling round with lumps of coke to steady it; then put in the copper, either blocks cut up into pieces of convenient size, or if this is not to be had, sheet copper doubled up; as the metal sinks down add more copper or old brass till the pot is nearly full of melted metal; now add the tin, and when this is melted and mixed, put in a piece or two of zinc; if this begins to flare add the rest of the zinc in, stir it well in, lift the pot off at

14*

once, skim the rubbish off the top, and pour into the mould. If, however, it does not flare up, put a little coal on to excite the fire, and cover over till it comes to a proper heat. As soon as the zinc begins to flare, add in the rest, and take the pot off the fire. If old brass alone is melted down no tin is required, but a small quantity of zinc. If part copper and part brass, add tin and zinc in proportion to the new copper, with a little extra zinc for the brass.

As soon as the boxes are run, it is the usual custom to open them at once, and to sprinkle the castings with water from the rose of a watering can, this has the effect of making them softer than they would otherwise be; the boxes are then emptied, and fresh moulds made while fresh metal is being melted.

When the casting is completed, draw the bearer forward, and let the bars all drop, so that the furnace can be effectually cleared from the clinkers, and put the pot among the ashes to cool gradually.

The moulding boxes may be of hard wood, well secured at the corners, either by dovetailing or by strong nails and iron corner plates, with guides to keep the boxes fair with one another. A few cross bars in the top box help to carry the sand.

Fresh green sand, the same as used for iron founding, mixed with a small quantity of coal dust, about one-twelfth part, should be sifted over the patterns on all sides to the thickness of about an inch, the box then filled up with old sand, and properly rammed up, and well pricked to let the air and gas escape, then remove the patterns, and dust over the mould with a little charcoal powder from a bag, or with a little flour, cover over the box again, and the mould is ready for pouring.

For long articles, spindles, bars, &c., make a good airhole at the opposite end from where the metal is poured, incline the box slightly, and pour the metal at the lower end; for flat, thin and straggling articles it is necessary to have two or more pouring holes, and to fill them all at the same time.

The pots generally used are the Stourbridge clay pots, and black lead pots, both kinds being made of various sizes up to 60 lbs.; the former are less durable, but much cheaper than the latter, they require to be carefully hardened by gradual exposure to the fire.

Clay pots are made of 2 parts raw Stourbridge clay to 1 of gas coke pulverized; well mixed up together with water, dried gently, and slightly baked in a kiln.

Black lead pots of 2 parts graphite, and 1 of fireclay, mixed with water, baked slightly in a kiln, but not completely until required for use.

The pots are made on a wood mould, the shape and size of the inside of the pot, the clay being plastered round it to the thickness desired.

ROPE.

To find the breaking Weight of an ordinary Tarred Hemp Rope.

(Circumference, ins.)$^2 \div 5 =$ Breaking weight, tons.

A rope should not be loaded with more than 1-3 its breaking weight.

To find Weight of Rope or Tarred Cordage.

(Circumference ins.)² × Length, ft. ÷ 24 = Weight, lbs.

Or,

(Circumference ins.)² ÷ 4 = Weight, lbs. per fathom.

To find Weight of Tarred Hawser or Manilla Rope.

(Circumference ins.)² ÷ 5 = Weight, lbs. per fathom.

To find Weight of Hawser-Laid Manilla.

(Circumference ins.)² ÷ 6 = Weight, lbs. per fathom.

WEIGHT.

To find the Weight of any Casting.

Width in ¼ ins. × Thickness in ⅛ ins., or vice versa, ÷ 10 × Length, ft. = Weight, lbs. cast iron.

For instance; to find the weight of a casting 3¼ ins. × 1⅛ ins. × 2 ft. 6 ins. long.

$$13 \times 9 \div 10 = 11.7 \times 2.5 = 29.25 \text{ lbs.}$$

This rule is very useful, and can easily be remembered in the following form.

Width in ¼ ins. × Thickness in ⅛ ins. or vice versa, cut off 1 figure for decimal, the result is lbs. per foot of length.

For wrought iron add 1-20th to the result; for lead add 1-2; for brass add 1-7th; for copper add 1-5th.

To find the Weight from the Areas.

Area, sq. ins. × Length, ft. × 3 1-7 = Weight, lbs. cast iron.

Multiplier for	Cast iron	3·156 or 3 1-7.
"	Wrought iron	3·312 or 3 1-3.
"	Lead	4·854
"	Brass	3·644
"	Copper	3·87

Or, Area, sq. ins. × 10 = lbs. per yard for wrought iron.

To find the Weight in cwts.

Area, sq. ins. × Length, ft. ÷ 31·9 = Weight, cwts. cast iron.

For wrought iron, divide by 33.6.

WEIGHT OF BOILER PLATES.

Thickness, ins.	$\frac{1}{16}$	$\frac{1}{8}$	$\frac{3}{16}$	$\frac{1}{4}$	$\frac{5}{16}$	$\frac{3}{8}$	$\frac{7}{16}$	$\frac{1}{2}$	$\frac{5}{8}$	$\frac{3}{4}$	$\frac{7}{8}$	1
Weight, lbs. per sq. ft.	2·5	5	7·5	10	12·5	15	17·5	20	25	30	35	40

For cast iron deduct 1-20th.

To find Weight of Boiler Plates in cwts.

$$\frac{\text{Area sq. ft.}}{\text{No. corresponding to thickness in table below.}} = \text{Weight, cwts.}$$

Thickness.	Divisor.	Thickness.	Divisor.	Thickness.	Divisor.
In.		In.		In.	
$\frac{1}{8}$	22·4	$\frac{3}{8}$	7·5	$\frac{5}{8}$	4·48
$\frac{3}{16}$	15·	$\frac{7}{16}$	6·3	$\frac{3}{4}$	3·73
$\frac{1}{4}$	11·2	$\frac{1}{2}$	5·6	$\frac{7}{8}$	3·2
$\frac{5}{16}$	9·	$\frac{9}{16}$	5·	1	2·8

CONTINUOUS CIRCULAR MOTION.

In mechanics, circular motion is transmitted by means of wheels, drums, or pulleys; and accordingly as the driving and driven are of equal or unequal diameters, so are equal or unequal velocities produced. Hence the principle on which the following rules are founded.

1. When time is not taken into Account.

Rule.—Divide the greater diameter, or number of teeth, by the lesser diameter or number of teeth; and the quotient is the number of revolutions the lesser will make, for one of the greater.

Example.—How many revolutions will a pinion of 20 teeth make, for 1 of a wheel with 125 ?

$$125 \div 20 = 6.25 \text{ or } 6\frac{1}{4} \text{ revolutions.}$$

To find the number of revolutions of the last, to one of the first, in a train of wheels and pinions.

Rule.—Divide the product of all the teeth in the driving by the product of all the teeth in the driven; and the quotient equal the ratio of velocity required.

Example 1.—Required the ratio of velocity of the last, to 1 of the first, in the following train of wheels and pinions; viz., pinions driving—the first of which contains 10 teeth, the second 15, and third 18. Wheels driven first, 15 teeth, second, 25, and third, 32.

$$\frac{10 \times 15 \times 18}{15 \times 25 \times 32} = \cdot 225 \text{ of a revolution the wheel will make to one of the pinion.}$$

Example 2.—A wheel of 42 teeth giving motion to one of 12, on which shaft is a pulley of 21 inches diameter driving one of 6; required the number of revolutions of the last pulley to one of the first wheel.

$$\frac{42 \times 21}{12 \times 6} = 12.25 \text{ or } 12\frac{1}{4} \text{ revolutions.}$$

Note.—Where increase or decrease of velocity is required to be communicated by wheel-work, it has been demonstrated that the number of teeth on each pinion should not be less than 1 to 6 of its wheel, unless there be some other important reason for a higher ratio.

2. *When Time must be regarded.*

RULE.—Multiply the diameter or number of teeth in the driver, by its velocity in any given time, and divide the product by the required velocity of the driven; the quotient equal the number of teeth or diameter of the driven, to produce the velocity required.

EXAMPLE 1.—If a wheel, containing 84 teeth, makes 20 revolutions per minute, how many must another contain, to work in contact, and make 60 revolutions in the same time ?

$$84 \times 20 \div 60 = 28 \text{ teeth.}$$

EXAMPLE 2.—From a shaft making 45 revolutions per minute, and with a pinion 9 inches diameter at the pitch line, I wish to transmit motion at 15 revolutions per minute; what, at the pitch line, must be the diameter of the wheel?

$$45 \times 9 \div 15 = 27 \text{ inches.}$$

EXAMPLE 3.—Required the diameter of a pulley to make 16 revolutions in the same time as one of 24 inches making 36.

$$24 \times 36 \div 16 = 54 \text{ inches.}$$

The distance between the centres and velocities of two wheels being given, to find their proper diameters.

RULE.—Divide the greatest velocity by the least; the quotient is the ratio of diameter the wheels must bear to each other.

Hence, divide the distance between the centres by the ratio + 1; the quotient equal the radius of the smaller wheel; and subtract the radius thus obtained from the distance between the centres; the remainder equal the radius of the other.

EXAMPLE.—The distance of two shafts from centre to centre is 50 inches, and the velocity of the one 25 revolutions per minute, the other is to make 80 in the same time; the proper diameters of the wheels at the pitch lines are required.

$80 \div 25 = 3.2$, ratio of velocity, and $50 \div 3.2 + 1 = 11.9$ the radius of the smaller wheel; then $50 - 11.9 = 38.1$, radius of larger; their diameters are $11.9 \times 2 = 23.8$ and $38.1 \times 2 = 76.2$ inches.

To obtain or diminish an accumulated velocity by means of wheels, pinions, or wheels, pinions, and pulleys, it is necessary that a proportional ratio of velocity should exist, and which is thus attained: multiply the given and required velocities together; and the square root of the product is the mean or proportionate velocity.

EXAMPLE.—Let the given velocity of a wheel containing 54 teeth equal 16 revolutions per minute, and the given diameter of an intermediate pulley equal 25 inches, to obtain a velocity of 81 revolutions in a machine; required the number of teeth in the intermediate wheel and diameter of the last pulley.

$\sqrt{81 \times 16} = 36$ mean velocity.

$54 \times 16 \div 36 = 24$ teeth and $25 \times 36 \div 81 = 11.1$ inches, diam. of pulley.

To determine the proportion of wheels for screw-cutting by a Lathe.

In a lathe properly adapted, screws to any degree of pitch, or number of threads in a given length, may be cut by means of a leading screw of any given pitch, accompanied with change wheels and pinions; coarse pitches being effected generally by means of one wheel and one pinion with a *carrier,* or *intermediate wheel,* which cause no variation or change of motion to take place. Hence the following

RULE.—Divide the number of threads in a given length of the screw which is to be cut, by the number of threads in the same length of the leading screw attached to the lathe ; and the quotient is the ratio that the wheel on the end of the screw must bear to that on the end of the lathe spindle.

EXAMPLE.—Let it be required to cut a screw with 5 threads in an inch, the leading screw being of $\frac{1}{2}$ inch pitch; or containing 2 threads in an inch ; what must be the ratio of wheels applied ?

$5 \div 2 = 2.5$, the ratio they must bear to each other.

Then suppose a pinion of 40 teeth be fixed upon for the spindle,—

$40 \times 2.5 = 100$ teeth for the wheel on the end of the screw.

But screws of a greater degree of fineness than about 8 threads in an inch are more conveniently cut by an additional wheel and pinion, because of the proper degree of velocity being more effectively attained ; and these, on account of revolving upon a stud, are commonly designated the *stud-wheels,* or *stud-wheel* and *pinion ;* but the mode of calculation and ratio of screw are the same as in the preceding rule. Hence, all that is further necessary is to fix upon any 3 wheels at pleasure, as those for the spindle and stud-wheels ; then multiply the number of teeth in the spindle-wheel by the ratio of the screw, and by the number of teeth in that wheel or pinion which is in contact with the wheel on the end of the screw ; divide the product by the stud-wheel in contact with the spindle-wheel ; and the quotient is the number of teeth required in the wheel on the end of the leading screw.

EXAMPLE.—Suppose a screw is required to be cut containing 25 threads in an inch, and the leading screw, as before, having two threads in an inch, and that a wheel of 60 teeth is fixed upon for the end of the spindle, 20 for the pinion in contact with the screw-wheel, and 100 for that in contact with the wheel on the end of the spindle ; required the number of teeth in the wheel for the end of the leading screw.

$$25 \div 2 = 12.5, \text{ and } \frac{60 \times 12.5 \times 20}{100} = 150 \text{ teeth.}$$

Or suppose the spindle and screw-wheels to be those fixed upon, also any one of the stud-wheels, to find the number of teeth in the other.

$$\frac{60 \times 12.5}{150 \times 100} = 20 \text{ teeth, or } \frac{60 \times 12.5 \times 20}{150} = 100 \text{ teeth.}$$

Table of Change Wheels for Screw-cutting; the leading Screw being ½ inch pitch, or containing 2 threads in an inch.

Number of threads in inch of screw.	Numb. of teeth in: Lathe spindle-wheel.	Numb. of teeth in: Leading screw-wheel.	Number of threads in inch of screw.	Number of teeth in: Lathe spindle-wheel.	Number of teeth in: Wheel in contact with spindle-wheel	Number of teeth in: Pinion in contact with screw-wheel.	Number of teeth in: Leading screw-wheel.	Number of threads in inch of screw.	Number of teeth in: Lathe spindle-wheel.	Number of teeth in: Wheel in contact with spindle-wheel	Number of teeth in: Pinion in contact with screw-wheel.	Number of teeth in: Leading screw-wheel.
1	80	40	8¼	40	55	20	60	19	50	95	20	100
1¼	80	50	8½	90	85	20	90	19½	80	120	20	130
1½	80	60	8¾	60	70	20	75	20	60	100	20	120
1¾	80	70	9½	90	90	20	95	20¼	40	90	20	90
2	80	90	9¾	40	60	20	65	21	80	120	20	140
2¼	80	90	10	60	75	20	80	22	60	110	20	120
2½	80	100	10½	50	70	20	75	22½	80	120	20	150
2¾	80	110	11	60	55	20	120	22¾	80	130	20	140
3	80	120	12	90	90	20	120	23¾	40	95	20	100
3¼	80	130	12¾	60	85	20	90	24	65	120	20	130
½	80	140	13	90	90	20	130	25	60	100	20	150
¾	80	150	13½	60	90	20	90	25½	30	85	20	90
4	40	80	13¾	80	100	20	110	26	70	130	20	140
4¼	40	85	14	90	90	20	140	27	40	90	20	120
4½	40	90	14¼	60	90	20	95	27½	40	100	20	110
4¾	40	95	15	90	90	20	150	28	75	140	20	150
5	40	100	16	60	80	20	120	28½	30	90	20	95
5½	40	110	16¼	80	100	20	130	30	70	140	20	150
6	40	120	16½	80	110	20	120	32	30	80	20	120
6½	40	130	17	45	85	20	90	33	40	110	20	120
7	40	140	17½	80	100	20	140	34	30	85	20	120
7½	40	150	18	40	60	20	120	35	60	140	20	150
8	30	120	18¾	80	100	20	150	36	30	90	20	120

Table by which to determine the Number of Teeth, or Pitch of Small Wheels, by what is commonly called the Manchester Principle.

Diametral Pitch.	Circular Pitch.	Diametral Pitch.	Circular Pitch.
3	1.047	9	.349
4	.785	10	.314
5	.628	12	.262
6	.524	14	.224
7	.449	16	.196
8	.393	20	.157

EXAMPLE 1.—Required the number of teeth that a wheel of 16 inches diameter will contain of a 10 pitch.

16 × 10 = 160 teeth, and the circular pitch = .314 inch.

EXAMPLE 2.—What must be the diameter of a wheel for a 9 pitch of 126 teeth ?

126 ÷ 9 = 14 inches diameter, circular pitch .349 inch.

NOTE.—The pitch is reckoned on the diameter of the wheel instead of the circumference, and designated wheels of 8 pitch, 12 pitch, &c.

Strength of the Teeth of Cast Iron Wheels at a given Velocity.

Pitch of teeth in inches.	Thickness of teeth in inches.	Breadth of teeth in inches.	Strength of teeth in horse-power at			
			3 feet per second.	4 feet per second.	6 feet per second.	8 feet per second.
3.99	1.9	7.6	20.57	27.43	41.14	54.85
3.78	1.8	7.2	17.49	23.32	34.98	46.64
3.57	1.7	6.8	14.73	19.65	29.46	39.28
3.36	1.6	6.4	12.28	16.38	24.56	32.74
3.15	1.5	6.	10.12	13.50	20.24	26.98
2.94	1.4	5.6	8.22	10.97	16.44	21.92
2.73	1.3	5.2	6.58	8.78	13.16	17.54
2.52	1.2	4.8	5.18	6.91	10.36	13.81
2.31	1.1	4.4	3.99	5.32	7.98	10.64
2.1	1.0	4.	3.00	4.00	6.00	8.00
1.89	.9	3.6	2.18	2.91	4.36	5.81
1.68	.8	3.2	1.53	2.04	.06	3.08
1.47	.7	2.8	1.027	1.37	2.04	2.72
1.26	.6	2.4	.64	.86	1.38	1.84
1.05	.5	2.	.375	.50	.75	1.00

WHEELS AND GUDGEONS.

To find size of Teeth necessary to transmit a given Horse Power.
(Tredgold.)

$$\frac{\text{Horse power} \times 240}{\text{Diameter of wheel, ft.} \times \text{Revs. per min.}} = \text{Strength of tooth.}$$

$$\sqrt{\frac{\text{Strength}}{\text{Breadth, ins.}}} = \text{Pitch, ins.} \qquad \frac{\text{Strength}}{(\text{Pitch, ins.})^2} = \text{Breadth, ins.}$$

The above rule will be found very suitable for a speed of circumference of about 240 feet per minute. For speeds above, add to 240 half the difference, for speeds below, deduct half the difference, between 240 and the actual speed, the result being a suitable multiplier.

For instance ; at 300 ft. per minute, 60 being the difference, 240 + 30 = 270 multiplier.

At 160 ft. per minute, 80 being the difference, 240 − 40 = 200 multiplier.

The reason being, that with higher speeds, the friction, wear, and liability to shocks is increased, at lower speeds decreased, and the teeth may advantageously be proportioned accordingly.

To find the Horse Power that any Wheel will transmit.

$$\frac{(\text{Pitch, ins.})^2 \times \text{Breadth, ins.} \times \text{Diameter ft.} \times \text{Revs. per minute}}{\text{Appropriate No. according to speed, as above.}} = \text{Horse Power.}$$

To find the multiplying number for any Wheel.

$$\frac{(\text{Pitch, ins.})^2 \times \text{Breadth, ins.} \times \text{Diameter ft.} \times \text{Revs. per minute}}{\text{Horse Power}} = \text{Multiplying No. as above.}$$

To find the size of Teeth to carry a given load in lbs.

Load, lbs. ÷ 1120 = Breaking strength of teeth.

Load, lbs. ÷ 280 = Strength for very low speeds, and for steady work; being 4 times the breaking strength.

Load, lbs. ÷ 140 = Strength for ordinary purposes of machinery; being 8 times the breaking strength.

Load, lbs. ÷ 100 = Strength for high speeds, and irregular work; or when the teeth are exposed to shocks.

As before,

$$\frac{\text{Strength}}{(\text{Pitch, ins.})^2} = \text{Breadth, ins.} \qquad \sqrt{\frac{\text{Strength}}{\text{Breadth, ins.}}} = \text{Pitch. ins.}$$

WATER.

To find the quantity of Water that will be discharged through an orifice, or pipe, in the side or bottom of a Vessel.

Area of orifice, sq. in. $\times$ $\begin{cases} \text{No. corresponding to height of surface} \\ \text{above orifice, as per table} \end{cases}$ = Cubic feet discharged per minute.

Height of Surface above Orifice.	Multiplier.	Height of Surface above Orifice.	Multiplier.	Height of Surface above Orifice.	Multiplier.
Ft.		Ft.		Ft.	
1	2·25	18	9·5	40	14·2
2	3·2	20	10·	45	15·1
4	4·5	22	10·5	50	16·
6	5·44	24	11·	60	17·4
8	6·4	26	11·5	70	18·8
10	7·1	28	12·	80	20·1
12	7·8	30	12·3	90	21·3
14	8·4	32	12·7	100	22·5
16	9·	35	13·3		

15

To find the size of hole necessary to discharge a given quantity of Water under a given head.

$$\frac{\text{Cubic ft. water discharged}}{\text{No. corresponding to height, as per table}} = \text{Area of orifice, sq. in.}$$

To find the height necessary to discharge a given quantity through a given orifice.

$$\frac{\text{Cubic ft. water discharged}}{\text{Area orifice, sq. inches.}} = \text{No. corresp. to height, as per table.}$$

The velocity of Water issuing from an orifice in the side or bottom of a vessel being ascertained to be as follows :

$$\sqrt{\text{Height ft. surface above orifice}} \times 5{\cdot}4 = \left\{ \begin{array}{c} \text{Velocity of water, ft.} \\ \text{per second.} \end{array} \right.$$

$$\sqrt{\text{Height ft.}} \times \text{Area orifice, ft.} \times 324 = \left\{ \begin{array}{c} \text{Cubic ft. discharged per} \\ \text{minute.} \end{array} \right.$$

$$\sqrt{\text{Height ft.}} \times \text{Area orifice, ins.} \times 2{\cdot}2 = \qquad \text{Do.} \qquad \text{Do.}$$

It may be observed, that the above rules represent the actual quantities that will be delivered through a hole cut in the plate ; if a short pipe be attached, the quantity will be increased, the greatest delivery with a straight pipe being attained with a length equal to 4 diameters, and being 1-3 more than the delivery through the plain hole ; the quantity gradually decreasing as the length of pipe is increased, till, with a length equal to 60 diameters the discharge again equals the discharge through the plain orifice. If a taper pipe be attached the delivery will be still greater, being 1½ times the delivery through the plain orifice ; and it is probable that if a pipe with curved decreasing taper were to be tried, the delivery through it would be equal to the theoretical discharge, which is about 1·65 the actual discharge through a plain hole.

To find the quantity of Water that will run through any orifice, the top of which is level with the surface of water as over a sluice or dam.

$$\sqrt{\left. \begin{array}{c} \text{Height, ft. from water surface to bot-} \\ \text{tom of orifice or top of dam} \end{array} \right\}} \times \left. \begin{array}{c} \text{Area of water} \\ \text{passage, sq. ft.} \end{array} \right\} \times 216$$

$$= \text{Cub. ft. discharged per minute.}$$

Or,

Two-thirds Area of water passage, sq. ins. $\times$ No. corresponding to height as per table, $=$ Cub. ft. discharged per minute.

To find the time in which a Vessel will empty itself through a given orifice.

$$\frac{\sqrt{\text{Height ft. surface above orifice}} \times \text{Area water surface, sq. ins.}}{\text{Area orifice, sq. in.} \times 3{\cdot}7}$$

$$= \text{Time required, seconds.}$$

The above rules are founded on Bank's experiments.

MECHANICAL TABLES

FOR THE USE OF

OPERATIVE SMITHS, MILLWRIGHTS,

AND

ENGINEERS.

MECHANICAL TABLES

FOR THE USE OF OPERATIVE SMITHS, MILLWRIGHTS, AND ENGINEERS.

The following Tables, originally dedicated to 'the National Association of the Forgers of Iron Work,' England, by JAMES FODEN, *will be found extremely useful to Smiths, generally, and are accompanied by Practical Examples.*—TEMPLETON.

DIAMETERS AND CIRCUMFERENCES OF CIRCLES.

Diam.	Circ.		Diam.	Circ.		Diam.		Circ.		Diam.		Circ.		Diam.		Circ.	
In.	Ft.	In.	In.	Ft.	In.	Ft.	In.	Ft.	In.	Ft.	In.	Ft.	In.	Ft.	In.	Ft.	In.
1	0	3⅛	5½	1	5¼	0	10	2	7⅜	1	2⅜	3	9⅛	1	6⅞	4	11¼
1⅛	0	3½	5⅝	1	5⅝					1	2½	3	9½	1	7	4	11⅝
1¼	0	3⅞	5¾	1	6	0	10⅛	2	7¾	1	2⅝	3	9⅞				
1⅜	0	4¼	5⅞	1	6⅜	0	10¼	2	8⅛	1	2¾	3	10¼	1	7⅛	5	0
1½	0	4⅝	6	1	6¾	0	10⅜	2	8½	1	2⅞	3	10⅜	1	7¼	5	0⅜
1⅝	0	5				0	10½	2	8⅞	1	3	3	11	1	7⅜	5	0⅞
1¾	0	5⅜	6⅛	1	7⅛	0	10⅝	2	9⅜					1	7½	5	1¼
1⅞	0	5⅞	6¼	1	7⅜	0	10¾	2	9¾	1	3⅛	3	11½	1	7⅝	5	1⅝
2	0	6¼	6⅜	1	8	0	10⅞	2	10⅛	1	3¼	3	11⅞	1	7¾	5	2
			6½	1	8⅜	0	11	2	10½	1	3⅜	4	0¼	1	7⅞	5	2⅜
2⅛	0	6⅝	6⅝	1	8¾					1	3½	4	0⅜	1	8	5	2¾
2¼	0	7	6¾	1	9⅛	0	11⅛	2	10⅞	1	3⅝	4	1				
2⅜	0	7⅜	6⅞	1	9½	0	11¼	2	11¼	1	3¾	4	1⅜	1	8⅛	5	3⅛
2½	0	7¾	7	1	9⅞	0	11⅜	2	11⅝	1	3⅞	4	1⅞	1	8¼	5	3⅝
2⅝	0	8¼				0	11½	3	0	1	4	4	2¼	1	8⅜	5	4
2¾	0	8⅝	7⅛	1	10⅜	0	11⅝	3	0½					1	8½	5	4⅜
2⅞	0	9	7¼	1	10¾	0	11¾	3	0⅞	1	4⅛	4	2⅝	1	8⅝	5	4¾
3	0	9⅜	7⅜	1	11⅛	0	11⅞	3	1¼	1	4¼	4	3	1	8¾	5	5⅛
			7½	1	11½	1	0	3	1⅜	1	4⅜	4	3⅜	1	8⅞	5	5½
3⅛	0	9¾	7⅝	1	11⅞					1	4½	4	3¾	1	9	5	5⅞
3¼	0	10⅛	7¾	2	0¼	1	0⅛	3	2	1	4⅝	4	4⅛				
3⅜	0	10½	7⅞	2	0⅝	1	0¼	3	2⅜	1	4¾	4	4½	1	9⅛	5	6⅜
3½	0	10⅞	8		1⅛	1	0⅜	3	2⅞	1	4⅞	4	5	1	9¼	5	6¾
3⅝	0	11⅜				1	0½	3	3¼	1	5	4	5⅜	1	9⅜	5	7⅛
3¾	0	11¾	8⅛	2	1½	1	0⅝	3	3⅝					1	9½	5	7½
3⅞	1	0⅛	8¼	2	1⅞	1	0¾	3	4	1	5⅛	4	5¾	1	9⅝	5	8
4	1	0½	8⅜	2	2¼	1	0⅞	3	4⅜	1	5¼	4	6⅛	1	9¾	5	8⅜
			8½	2	2⅝	1	1	3	4¾	1	5⅜	4	6½	1	9⅞	5	8¾
4⅛	1	0⅞	8⅝	2	3					1	5½	4	6⅞	1	10	5	9
4¼	1	1¼	8¾	2	3⅜	1	1⅛	3	5⅛	1	5⅝	4	7⅜				
4⅜	1	1⅝	8⅞	2	3⅞	1	1¼	3	5⅝	1	5¾	4	7¾	1	10⅛	5	9½
4½	1	2⅛	9	2	4¼	1	1⅜	3	6	1	5⅞	4	8⅛	1	10¼	5	9⅞
4⅝	1	2½				1	1½	3	6⅜	1	6	4	8½	1	10⅜	5	10¼
4¾	1	2⅞	9⅛	2	4⅝	1	1⅝	3	6¾					1	10½	5	10⅝
4⅞	1	3¼	9¼	2	5	1	1¾	3	7⅛	1	6⅛	4	8⅞	1	10⅝	5	11
5	1	3⅝	9⅜	2	5⅜	1	1⅞	3	7½	1	6¼	4	9¼	1	10¾	5	11⅜
			9½	2	5¾	1	2	3	7⅞	1	6⅜	4	9⅝	1	10⅞	5	11¾
5⅛	1	4	9⅝	2	6⅛					1	6½	4	10	1	11	6	0¼
5¼	1	4⅜	9¾	2	6⅝	1	2⅛	3	8⅜	1	6⅝	4	10⅜				
5⅜	1	4⅞	9⅞	2	7	1	2¼	3	8¾	1	6¾	4	10⅞	1	11⅛	6	0⅝

Diam.		Circ.		Diam.		Circ.		Diam.		Circ.		Diam.		Circ.		Diam.		Circ.	
Ft.	In.	Ft.	In.	Ft.	In.	Ft.	In.	Ft.	In.	Ft.	In.	Ft.	In.	Ft.	In.	Ft.	In.	Ft	In.
1	11¼	6	1	2	4⅞	7	6⅝	2	10⅜	8	11⅞	3	4	10	5⅝	3	9½	11	10⅞
1	11⅜	6	1⅜	2	5	7	7	2	10½	9	0⅜					3	9⅝	11	11¼
1	11½	6	1¾					2	10⅝	9	0¾	3	4⅛	10	6	3	9¾	11	11⅝
1	11⅝	6	2⅛	2	5⅛	7	7½	2	10¾	9	1⅛	3	4¼	10	6⅜	3	9⅞	12	0
1	11¾	6	2½	2	5¼	7	7⅞	2	10⅞	9	1½	3	4⅜	10	6¾	3	10	12	0½
1	11⅞	6	3	2	5⅜	7	8¼	2	11	9	1⅞	3	4½	10	7⅛				
2	0	6	3⅜	2	5½	7	8⅝					3	4⅝	10	7½	3	10⅛	12	0⅞
				2	5⅝	7	9	2	11⅛	9	2¼	3	4¾	10	8	3	10¼	12	1¼
2	0⅛	6	3¾	2	5¾	7	9⅜	2	11¼	9	2⅝	3	4⅞	10	8⅜	3	10⅜	12	1⅝
2	0¼	6	4⅛	2	5⅞	7	9¾	2	11⅜	9	3	3	5	10	8¾	3	10½	12	2
2	0⅜	6	4½	2	6	7	10⅛	2	11½	9	3½					3	10⅝	12	2⅜
2	0½	6	4⅞					2	11⅝	9	3⅞	3	5⅛	10	9⅛	3	10¾	12	2¾
2	0⅝	6	5¼	2	6⅛	7	10⅝	2	11¾	9	4¼	3	5¼	10	9½	3	10⅞	12	3¼
2	0¾	6	5¾	2	6¼	7	11	2	11⅞	9	4⅝	3	5⅜	10	9⅞	3	11	12	3⅝
2	0⅞	6	6⅛	2	6⅜	7	11⅜	3	0	9	5	3	5½	10	10¼				
2	1	6	6½	2	6½	7	11¾					3	5⅝	10	10¾	3	11⅛	12	4
				2	6⅝	8	0⅛	3	0⅛	9	5⅜	3	5¾	10	11⅛	3	11¼	12	4⅜
2	1⅛	6	6⅞	2	6¾	8	0½	3	0¼	9	5⅞	3	5⅞	10	11½	3	11⅜	12	4¾
2	1¼	6	7¼	2	6⅞	8	0⅞	3	0⅜	9	6¼	3	6	10	11⅞	3	11½	12	5⅛
2	1⅜	6	7⅝	2	7	8	1¼	3	0½	9	6⅝					3	11⅝	12	5½
2	1½	6	8					3	0⅝	9	7	3	6⅛	11	0¼	3	11¾	12	6
2	1⅝	6	8½	2	7⅛	8	1¾	3	0¾	9	7⅜	3	6¼	11	0⅝	3	11⅞	12	6⅜
2	1¾	6	8⅞	2	7¼	8	2⅛	3	0⅞	9	7¾	3	6⅜	11	1⅛	4	0	12	6¾
2	1⅞	6	9¼	2	7⅜	8	2½	3	1	9	8⅛	3	6½	11	1½				
2	2	6	9⅝	2	7½	8	2⅞					3	6⅝	11	1⅞	4	0⅛	12	7⅛
				2	7⅝	8	3¼	3	1⅛	9	8⅝	3	6¾	11	2¼	4	0¼	12	7½
2	2⅛	6	10	2	7¾	8	3¾	3	1¼	9	9	3	6⅞	11	2⅝	4	0⅜	12	7⅞
2	2¼	6	10⅜	2	7⅞	8	4⅛	3	1⅜	9	9⅜	3	7	11	3	4	0½	12	8¼
2	2⅜	6	10¾	2	8	8	4½	3	1½	9	9¾					4	0⅝	12	8¾
2	2½	6	11⅛					3	1⅝	9	10⅛	3	7⅛	11	3⅜	4	0¾	12	9⅛
2	2⅝	6	11⅝	2	8⅛	8	4⅞	3	1¾	9	10½	3	7¼	11	3⅞	4	0⅞	12	9½
2	2¾	7	0	2	8¼	8	5¼	3	1⅞	9	10⅞	3	7⅜	11	4¼	4	1	12	9⅞
2	2⅞	7	0⅜	2	8⅜	8	5⅝	3	2	9	11⅜	3	7½	11	4⅝				
2	3	7	0¾	2	8½	8	6					3	7⅝	11	5	4	1⅛	12	10¼
				2	8⅝	8	6⅜	3	2⅛	9	11¾	3	7¾	11	5⅜	4	1¼	12	10⅝
2	3⅛	7	1⅛	2	8¾	8	6⅞	3	2¼	10	0⅛	3	7⅞	11	5¾	4	1⅜	12	11
2	3¼	7	1½	2	8⅞	8	7¼	3	2⅜	10	0½	3	8	11	6⅛	4	1½	12	11½
2	3⅜	7	2	2	9	8	7⅝	3	2½	10	0⅞					4	1⅝	12	11⅞
2	3½	7	2⅜					3	2⅝	10	1¼	3	8⅛	11	6½	4	1¾	13	0¼
2	3⅝	7	2¾	2	9⅛	8	8	3	2¾	10	1⅝	3	8¼	11	7	4	1⅞	13	0⅝
2	3¾	7	3⅛	2	9¼	8	8⅜	3	2⅞	10	2⅛	3	8⅜	11	7⅜	4	2	13	1
2	3⅞	7	3½	2	9⅜	8	8¾	3	3	10	2½	3	8½	11	7¾				
2	4	7	3⅞	2	9½	8	9¼					3	8⅝	11	8⅛	4	2⅛	13	1½
				2	9⅝	8	9⅝	3	3⅛	10	2⅞	3	8¾	11	8½	4	2¼	13	1⅞
2	4⅛	7	4¼	2	9¾	8	10	3	3¼	10	3¼	3	8⅞	11	8⅞	4	2⅜	13	2¼
2	4¼	7	4¾	2	9⅞	8	10⅜	3	3⅜	10	3⅝	3	9	11	9¼	4	2½	13	2⅝
2	4⅜	7	5⅛	2	10	8	10¾	3	3½	10	4					4	2⅝	13	3
2	4½	7	5½					3	3⅝	10	4½	3	9⅛	11	9¾	4	2¾	13	3⅜
2	4⅝	7	5⅞	2	10⅛	8	11⅛	3	3¾	10	4⅞	3	9¼	11	10⅛	4	2⅞	13	3¾
2	4¾	7	6¼	2	10¼	8	11½	3	3⅞	10	5¼	3	9⅜	11	10½	4	3	13	4⅛

Diam.		Circ.		Diam.		Circ.		Diam.		Circ.		Diam.		Circ.		Diam.		Circ.	
Ft.	In.	Ft.	In.	Ft.	In.	Ft.	In.	Ft.	in.	Ft.	In.	Ft.	In.	Ft.	In.	Ft.	In.	Ft.	In.
4	3 1/8	13	4 1/2	4	8 3/4	14	10 1/4	5	2 1/4	16	3 1/2	5	7 7/8	17	9 1/8	6	1 3/8	19	2 1/2
4	3 1/4	13	5	4	8 7/8	14	10 5/8	5	2 3/8	16	3 7/8	5	8	17	9 5/8	6	1 1/2	19	2 7/8
4	3 3/8	13	5 3/8	4	9	14	11	5	2 1/2	16	4 1/4					6	1 5/8	19	3 1/4
4	3 1/2	13	5 3/4					5	2 5/8	16	4 5/8	5	8 1/8	17	10	6	1 3/4	19	3 5/8
4	3 5/8	13	6 1/8	4	9 1/8	14	11 3/8	5	2 3/4	16	5 1/8	5	8 1/4	17	10 3/8	6	1 7/8	19	4
4	3 3/4	13	6 1/2	4	9 1/4	14	11 3/4	5	2 7/8	16	5 1/2	5	8 3/8	17	10 3/4	6	2	19	4 3/8
4	3 7/8	13	6 7/8	4	9 3/8	15	0 1/8	5	3	16	5 7/8	5	8 1/2	17	11 1/8				
4	4	13	7 1/4	4	9 1/2	15	0 5/8					5	8 5/8	17	11 1/2	6	2 1/8	19	4 3/4
				4	9 5/8	15	1	5	3 1/8	16	6 1/4	5	8 3/4	17	11 7/8	6	2 1/4	19	5 1/4
4	4 1/8	13	7 3/4	4	9 3/4	15	1 3/8	5	3 1/4	16	6 5/8	5	8 7/8	18	0 3/8	6	2 3/8	19	5 5/8
4	4 1/4	13	8 1/8	4	9 7/8	15	1 3/4	5	3 3/8	16	7	5	9	18	0 3/4	6	2 1/2	19	6
4	4 3/8	13	8 1/2	4	10	15	2 1/8	5	3 1/2	16	7 3/8					6	2 5/8	19	6 3/8
4	4 1/2	13	8 7/8					5	3 5/8	16	7 7/8	5	9 1/8	18	1 1/8	6	2 3/4	19	6 3/4
4	4 5/8	13	9 1/4	4	10 1/8	15	2 1/2	5	3 3/4	16	8 1/4	5	9 1/4	18	1 1/2	6	2 7/8	19	7 1/8
4	4 3/4	13	9 5/8	4	10 1/4	15	2 7/8	5	3 7/8	16	8 5/8	5	9 3/8	18	1 7/8	6	3	19	7 1/2
4	4 7/8	13	10	4	10 3/8	15	3 3/8	5	4	16	9	5	9 1/2	18	2 1/4				
4	5	13	10 1/2	4	10 1/2	15	3 3/4					5	9 5/8	18	2 5/8	6	3 1/8	19	8
				4	10 5/8	15	4 1/8	5	4 1/8	16	9 5/8	5	9 3/4	18	3	6	3 1/4	19	8 3/8
4	5 1/8	13	10 7/8	4	10 3/4	15	4 1/2	5	4 1/4	16	9 3/4	5	9 7/8	18	3 1/2	6	3 3/8	19	8 3/4
4	5 1/4	13	11 1/4	4	10 7/8	15	4 7/8	5	4 3/8	16	10 1/8	5	10	18	3 7/8	6	3 1/2	19	9 1/8
4	5 3/8	13	11 5/8	4	11	15	5 1/4	5	4 1/2	16	10 5/8					6	3 5/8	19	9 1/2
4	5 1/2	14	0					5	4 5/8	16	11	5	10 1/8	18	4 1/4	6	3 3/4	19	9 7/8
4	5 5/8	14	0 3/8	4	11 1/8	15	5 5/8	5	4 3/4	16	11 3/8	5	10 1/4	18	4 5/8	6	3 7/8	19	10 1/4
4	5 3/4	14	0 7/8	4	11 1/4	15	6 1/8	5	4 7/8	16	11 3/4	5	10 3/8	18	5	6	4	19	10 3/4
4	5 7/8	14	1 1/4	4	11 3/8	15	6 1/2	5	5	17	0 1/8	5	10 1/2	18	5 3/8				
4	6	14	1 5/8	4	11 1/2	15	6 7/8					5	10 5/8	18	5 7/8	6	4 1/8	19	11 1/8
				4	11 5/8	15	7 1/4	5	5 1/8	17	0 5/8	5	10 3/4	18	6 1/4	6	4 1/4	19	11 1/2
4	6 1/8	14	2	4	11 3/4	15	7 5/8	5	5 1/4	17	1	5	10 7/8	18	6 5/8	6	4 3/8	19	11 7/8
4	6 1/4	14	2 3/8	4	11 7/8	15	8	5	5 3/8	17	1 3/8	5	11	18	7	6	4 1/2	20	0 1/4
4	6 3/8	14	2 3/4	5	0	15	8 3/8	5	5 1/2	17	1 3/4					6	4 5/8	20	0 5/8
4	6 1/2	14	3 1/8					5	5 5/8	17	2 1/8	5	11 1/8	18	7 3/8	6	4 3/4	20	1
4	6 5/8	14	3 1/2	5	0 1/8	15	8 7/8	5	5 3/4	17	2 1/2	5	11 1/4	18	7 [illegible]	6	4 7/8	20	1 3/8
4	6 3/4	14	4	5	0 1/4	15	9 1/4	5	5 7/8	17	2 7/8	5	11 3/8	18	8 1/8	6	5	20	1 7/8
4	6 7/8	14	4 3/8	5	0 3/8	15	9 5/8	5	6	17	3 1/4	5	11 1/2	18	8 1/2				
4	7	14	4 3/4	5	0 1/2	15	10					5	11 5/8	18	9	6	5 1/8	20	2 1/4
				5	0 5/8	15	10 3/8	5	6 1/8	17	3 5/8	5	11 3/4	18	9 3/8	6	5 1/4	20	2 5/8
4	7 1/8	14	5 1/8	5	0 3/4	15	10 3/4	5	6 1/4	17	4 1/8	5	11 7/8	18	9 3/4	6	5 3/8	20	3
4	7 1/4	14	5 1/2	5	0 7/8	15	11 1/8	5	6 3/8	17	4 1/2	6	0	18	10 1/8	6	5 1/2	20	3 3/8
4	7 3/8	14	5 7/8	5	1	15	11 5/8	5	6 1/2	17	4 7/8					6	5 5/8	20	3 3/4
4	7 1/2	14	6 1/4					5	6 5/8	17	5 1/4	6	0 1/8	18	10 1/2	6	5 3/4	20	4 1/4
4	7 5/8	14	6 3/4	5	1 1/8	16	0	5	6 3/4	17	5 5/8	6	0 1/4	18	10 7/8	6	5 7/8	20	4 5/8
4	7 3/4	14	7 1/8	5	1 1/4	16	0 3/8	5	6 7/8	17	6	6	0 3/8	18	11 1/4	6	6	20	5
4	7 7/8	14	7 1/2	5	1 3/8	16	0 3/4	5	7	17	6 3/8	6	0 1/2	18	11 3/4				
4	8	14	7 7/8	5	1 1/2	16	1 1/8					6	0 5/8	19	0 1/8	6	6	20	5 3/4
				5	1 5/8	16	1 1/2	5	7 1/8	17	6 7/8	6	0 3/4	19	0 1/2	6	6 1/4	20	5 7/8
4	8 1/8	14	8 1/4	5	1 3/4	16	1 7/8	5	7 1/4	17	7 1/4	6	0 7/8	19	0 7/8	6	6 3/8	20	6 1/4
4	8 1/4	14	8 5/8	5	1 7/8	16	2 3/8	5	7 3/8	17	7 5/8	6	1	19	1 1/4	6	6 1/2	20	6 1/2
4	8 3/8	14	9	5	2	16	2 3/4	5	7 1/2	17	8					6	[illegible]	20	7
4	8 1/2	14	9 1/2					5	7 5/8	17	8 3/8	6	1 1/8	19	1 5/8	6	6 3/4	20	7 3/8
4	8 5/8	14	9 7/8	5	2 1/8	16	3 1/8	5	7 3/4	17	8 3/4	6	1 1/4	19	2	6	6 7/8	20	7 [illegible]

Diam.		Circ.		Diam.		Circ.		Diam.		Circ.		Diam.		Circ.		Diam.		Circ.	
Ft.	In.	Ft.	In.	Ft.	In.	Ft.	In.	Ft.	In.	Ft.	In.	Ft.	In.	Ft.	In.	Ft.	In.	Ft.	In.
6	7	20	8 1/8	7	0 1/2	22	1 3/8	7	6 1/8	23	7 1/8	7	11 3/4	25	0 3/4	8	5 1/4	26	6
				7	0 5/8	22	1 7/8	7	6 1/4	23	7 1/2	7	11 7/8	25	1 1/8	8	5 3/8	26	6 3/8
6	7 1/8	20	8 1/2	7	0 3/4	22	2 1/4	7	6 3/8	23	7 7/8	8	0	25	1 1/2	8	5 1/2	26	6 3/4
6	7 1/4	20	8 7/8	7	0 7/8	22	2 5/8	7	6 1/2	23	8 1/4					8	5 5/8	26	7 1/4
6	7 3/8	20	9 1/4	7	1	22	3	7	6 5/8	23	8 5/8	8	0 1/8	25	1 7/8	8	5 3/4	26	7 5/8
6	7 1/2	20	9 3/4					7	6 3/4	23	9	8	0 1/4	25	2 1/4	8	5 7/8	26	8
6	7 5/8	20	10 1/8	7	1 1/8	22	3 3/8	7	6 7/8	23	9 3/8	8	0 3/8	25	2 3/4	8	6	26	8 3/8
6	7 3/4	20	10 1/2	7	1 1/4	22	3 3/4	7	7	23	9 7/8	8	0 1/2	25	3 1/8				
6	7 7/8	20	10 7/8	7	1 3/8	22	4 1/8					8	0 5/8	25	3 1/2	8	6 1/8	26	8 3/4
6	8	20	11 1/4	7	1 1/2	22	4 1/2	7	7 1/8	23	10 1/4	8	0 3/4	25	3 7/8	8	6 1/4	26	9 1/8
				7	1 5/8	22	5	7	7 1/4	23	10 5/8	8	0 7/8	25	4 1/4	8	6 3/8	26	9 1/2
6	8 1/8	20	11 5/8	7	1 3/4	22	5 3/8	7	7 3/8	23	11	8	1	25	4 5/8	8	6 1/2	26	10
6	8 1/4	21	0	7	1 7/8	22	5 3/4	7	7 1/2	23	11 3/8					8	6 5/8	26	10 3/8
6	8 3/8	21	0 1/2	7	2	22	6 1/8	7	7 5/8	23	11 3/4	8	1 1/8	25	5 1/8	8	6 3/4	26	10 3/4
6	8 1/2	21	0 7/8					7	7 3/4	24	0 1/8	8	1 1/4	25	5 1/2	8	6 7/8	26	11 1/8
6	8 5/8	21	1 1/4	7	2 1/8	22	6 1/2	7	7 7/8	24	0 5/8	8	1 3/8	25	5 7/8	8	7	26	11 1/2
6	8 3/4	21	1 5/8	7	2 1/4	22	6 7/8	7	8	24	1	8	1 1/2	25	6 1/4				
6	8 7/8	21	2	7	2 3/8	22	7 1/4					8	1 5/8	25	6 5/8	8	7 1/8	26	11 7/8
6	9	21	2 3/8	7	2 1/2	22	7 5/8	7	8 1/8	24	1 3/8	8	1 3/4	25	7	8	7 1/4	27	0 1/4
				7	2 5/8	22	8 1/8	7	8 1/4	24	1 3/4	8	1 7/8	25	7 3/8	8	7 3/8	27	0 5/8
6	9 1/8	21	2 7/8	7	2 3/4	22	8 1/2	7	8 3/8	24	2 1/8	8	2	25	7 7/8	8	7 1/2	27	1 1/8
6	9 1/4	21	3 1/4	7	2 7/8	22	8 7/8	7	8 1/2	24	2 1/2					8	7 5/8	27	1 1/2
6	9 3/8	21	3 5/8	7	3	22	9 1/4	7	8 5/8	24	2 7/8	8	2 1/8	25	8 1/4	8	7 3/4	27	1 7/8
6	9 1/2	21	4					7	8 3/4	24	3 3/8	8	2 1/4	25	8 5/8	8	7 7/8	27	2 1/4
6	9 5/8	21	4 3/8	7	3 1/8	22	9 5/8	7	8 7/8	24	3 3/4	8	2 3/8	25	9	8	8	27	2 5/8
6	9 3/4	21	4 3/4	7	3 1/4	22	10	7	9	24	4 1/8	8	2 1/2	25	9 3/8				
6	9 7/8	21	5 1/8	7	3 3/8	22	10 3/8					8	2 5/8	25	9 3/4	8	8 1/8	27	3
6	10	21	5 1/2	7	3 1/2	22	10 7/8	7	9 1/8	24	4 1/2	8	2 3/4	25	10 1/8	8	8 1/4	27	3 1/2
				7	3 5/8	22	11 1/4	7	9 1/4	24	4 7/8	8	2 7/8	25	10 1/2	8	8 3/8	27	3 7/8
6	10 1/8	21	6	7	3 3/4	22	11 5/8	7	9 3/8	24	5 1/4	8	3	25	11	8	8 1/2	27	4 1/4
6	10 1/4	21	6 3/8	7	3 7/8	23	0	7	9 1/2	24	5 5/8					8	8 5/8	27	4 5/8
6	10 3/8	21	6 3/4	7	4	23	0 3/8	7	9 5/8	24	6 1/8	8	3 1/8	25	11 3/8	8	8 3/4	27	5
6	10 1/2	21	7 1/8					7	9 3/4	24	6 1/2	8	3 1/4	25	11 3/4	8	8 7/8	27	5 3/8
6	10 5/8	21	7 1/2	7	4 1/8	23	0 3/4	7	9 7/8	24	6 7/8	8	3 3/8	26	0 1/8	8	9	27	5 3/4
6	10 3/4	21	7 7/8	7	4 1/4	23	1 1/8	7	10	24	7 1/4	8	3 1/2	26	0 1/2				
6	10 7/8	21	8 1/4	7	4 3/8	23	1 5/8					8	3 5/8	26	0 7/8	8	9 1/8	27	6 1/4
6	11	21	8 3/4	7	4 1/2	23	2	7	10 1/8	24	7 5/8	8	3 3/4	26	1 3/8	8	9 1/4	27	6 5/8
				7	4 5/8	23	2 3/8	7	10 1/4	24	8	8	3 7/8	26	1 3/4	8	9 3/8	27	7
6	11 1/8	21	9 1/8	7	4 3/4	23	2 3/4	7	10 3/8	24	8 3/8	8	4	26	2 1/8	8	9 1/2	27	7 3/8
6	11 1/4	21	9 1/2	7	4 7/8	23	3 1/8	7	10 1/2	24	8 7/8					8	9 5/8	27	7 3/4
6	11 3/8	21	9 7/8	7	5	23	3 1/2	7	10 5/8	24	9 1/4	8	4 1/8	26	2 1/2	8	9 3/4	27	8 1/4
6	11 1/2	21	10 1/4					7	10 3/4	24	9 5/8	8	4 1/4	26	2 7/8	8	9 7/8	27	8 1/2
6	11 5/8	21	10 5/8	7	5 1/8	23	3 7/8	7	10 7/8	24	10	8	4 3/8	26	3 1/4	8	10	27	9
6	11 3/4	21	11	7	5 1/4	23	4 3/8	7	11	24	10 3/8	8	4 1/2	26	3 5/8				
6	11 7/8	21	11 1/2	7	5 3/8	23	4 3/4					8	4 5/8	26	4 1/8	8	10 1/8	27	9 3/8
7	0	21	11 7/8	7	5 1/2	23	5 1/8	7	11 1/8	24	10 3/4	8	4 3/4	26	4 1/2	8	10 1/4	27	9 3/4
				7	5 5/8	23	5 1/2	7	11 1/4	24	11 1/8	8	4 7/8	26	4 7/8	8	10 3/8	27	10 1/4
7	0 1/8	22	0 1/4	7	5 3/4	23	5 7/8	7	11 3/8	24	11 1/2	8	5	26	5 1/4	8	10 1/2	27	10 5/8
7	0 1/4	22	0 5/8	7	5 7/8	23	6 1/4	7	11 1/2	25	0					8	10 5/8	27	10 7/8
7	0 3/8	22	1	7	6	23	6 5/8	7	11 5/8	25	0 3/8	8	5 1/8	26	5 5/8	8	10 3/4	27	11 1/4

Diam.		Circ.		Dia.		Circ.		Diam.		Circ.		Diam.		Circ.		Diam.		Circ.	
Ft.	In.	Ft.	In.	F.	I.	Ft.	In.	Ft.	In.	Ft.	In	Ft.	In.	Ft.	In.	Ft.	In.	Ft.	In.
8	10⅞	27	11¾	9	4⅜	29	5	9	10	30	10⅝	10	3½	32	3⅞	10	9⅛	33	9⅝
8	11	28	0⅛	9	4½	29	5⅜					10	3⅝	32	4⅜	10	9¼	33	10
				9	4⅝	29	5¾	9	10⅛	30	11	10	3¾	32	4¾	10	9⅜	33	10⅜
8	11⅛	28	0½	9	4¾	29	6⅛	9	10¼	30	11⅜	10	3⅞	32	5⅛	10	9½	33	10¾
8	11¼	28	0⅞	9	4⅞	29	6½	9	10⅜	30	11⅞	10	4	32	5½	10	9⅝	33	11⅛
8	11⅜	28	1¼	9	5	29	7	9	10½	31	0¼					10	9¾	33	11½
8	11½	28	1⅝					9	10⅝	31	0⅝	10	4⅛	32	5⅞	10	9⅞	34	0
8	11⅝	28	2	9	5⅛	29	7⅜	9	10¾	31	1	10	4¼	32	6¼	10	10	34	0⅜
8	11¾	28	2½	9	5¼	29	7¾	9	10⅞	31	1⅜	10	4⅜	32	6⅝				
8	11⅞	28	2⅞	9	5⅜	29	8⅛	9	11	31	1¾	10	4½	32	7	10	10⅛	34	0¾
9	0	28	3¼	9	5½	29	8½					10	4⅝	32	7½	10	10¼	34	1⅛
				9	5⅝	29	8⅞	9	11⅛	31	2¼	10	4¾	32	7⅞	10	10⅜	34	1½
9	0⅛	28	3⅝	9	5¾	29	9¼	9	11¼	31	2⅝	10	4⅞	32	8¼	10	10½	34	1⅞
9	0¼	28	4	9	5⅞	29	9¾	9	11⅜	31	3	10	5	32	8⅝	10	10⅝	34	2⅜
9	0⅜	28	4⅜	9	6	29	10⅛	9	11½	31	3⅜					10	10¾	34	2¾
9	0½	28	4¾					9	11⅝	31	3¾	10	5⅛	32	9	10	10⅞	34	3⅛
9	0⅝	28	5¼	9	6⅛	29	10½	9	11¾	31	4⅛	10	5¼	32	9⅜	10	11	34	3½
9	0¾	28	5⅝	9	6¼	29	10⅞	9	11⅞	31	4½	10	5⅜	32	9⅞				
9	0⅞	28	6	9	6⅜	29	11¼					10	5½	32	10¼	10	11⅛	34	3⅞
9	1	28	6⅜	9	6½	29	11⅝	10	0	31	4⅞	10	5⅝	32	10⅝	10	11¼	34	4¼
				9	6⅝	30	0	10	0⅛	31	5⅜	10	5¾	32	11	10	11⅜	34	4⅝
9	1⅛	28	6¾	9	6¾	30	0½	10	0¼	31	5¾	10	5⅞	32	11⅜	10	11½	34	5
9	1¼	28	7⅛	9	6⅞	30	0⅞	10	0⅜	31	6⅛	10	6	32	11¾	10	11⅝	34	5½
9	1⅜	28	7½	9	7	30	1¼	10	0½	31	6½					10	11¾	34	5⅞
9	1½	28	8					10	0⅝	31	6⅞	10	6⅛	33	0⅛	10	11⅞	34	6¼
9	1⅝	28	8⅜	9	7⅛	30	1⅝	10	0¾	31	7¼	10	6¼	33	0⅝	11	0	34	6⅝
9	1¾	28	8¾	9	7¼	30	2	10	0⅞	31	7⅝	10	6⅜	33	1				
9	1⅞	28	9⅛	9	7⅜	30	2⅜	10	1	31	8⅛	10	6½	33	1⅜	11	0⅛	34	7
9	2	28	9½	9	7½	30	2¾					10	6⅝	33	1¾	11	0¼	34	7⅜
				9	7⅝	30	3⅛	10	1⅛	31	8½	10	6¾	33	2⅛	11	0⅜	34	7¾
9	2⅛	28	9⅞	9	7¾	30	3⅝	10	1¼	31	8⅞	10	6⅞	33	2½	11	0½	34	8¼
9	2¼	28	10¼	9	7⅞	30	4	10	1⅜	31	9¼	10	7	33	2⅞	11	0⅝	34	8⅝
9	2⅜	28	10¾	9	8	30	4⅜	10	1½	31	9⅝					11	0¾	34	9
9	2½	28	11⅛					10	1⅝	31	10	10	7⅛	33	3⅜	11	0⅞	34	9⅜
9	2⅝	28	11½	9	8⅛	30	4¾	10	1¾	31	10⅜	10	7¼	33	3¾	11	1	34	9¾
9	2¾	28	11⅞	9	8¼	30	5⅛	10	1⅞	31	10⅞	10	7⅜	33	4⅛				
9	2⅞	29	0¼	9	8⅜	30	5½	10	2	31	11¼	10	7½	33	4½	11	1⅛	34	10⅛
9	3	29	0⅝	9	8½	30	6					10	7⅝	33	4⅞	11	1¼	34	10½
				9	8⅝	30	6⅜	10	2⅛	31	11⅝	10	7¾	33	5¼	11	1⅜	34	11
9	3⅛	29	1	9	8¾	30	6¾	10	2¼	32	0	10	7⅞	33	5⅝	11	1½	34	11⅜
9	3¼	29	1½	9	8⅞	30	7⅛	10	2⅜	32	0⅜	10	8	33	6⅛	11	1⅝	34	11¾
9	3⅜	29	1⅞	9	9	30	7½	10	2½	32	0¾					11	1¾	35	0⅛
9	3½	29	2¼					10	2⅝	32	1⅛	10	8⅛	33	6½	11	1⅞	35	0½
9	3⅝	29	2⅝	9	9⅛	30	7⅞	10	2¾	32	1½	10	8¼	33	6⅞	11	2	35	0⅞
9	3¾	29	3	9	9¼	30	8¼	10	2⅞	32	2	10	8⅜	33	7¼				
9	3⅞	29	3⅜	9	9⅜	30	8⅝	10	3	32	2⅜	10	8½	33	7⅝	11	2⅛	35	1¼
9	4	29	3¾	9	9½	30	9⅛					10	8⅝	33	8	11	2¼	35	1¾
				9	9⅝	30	9½	10	3⅛	32	2¾	10	8¾	33	8⅜	11	2⅜	35	2⅛
9	4⅛	29	4¼	9	9¾	30	9⅞	10	3¼	32	3⅛	10	8⅞	33	8¾	11	2½	35	2½
9	4¼	29	4⅝	9	9⅞	30	10¼	10	3⅜	32	3½	10	9	33	9¼	11	2⅝	35	2⅞

Diam.		Circum.		Diam.		Circum.		Diam.		Circum.		Diam.		Circum	
Ft	In.	Ft.	In.	Ft.	In.	Ft.	In.	Ft.	In.	Ft.	In.	Ft.	In.	Ft.	In
11	2¾	35	3¼	11	5¾	36	0¾	11	8¾	36	10⅛	11	11¾	37	7½
11	2⅞	35	3⅝	11	5⅞	36	1⅛	11	8⅞	36	10½	11	11⅞	37	7⅞
11	3	35	4	11	6	36	1½	11	9	36	10⅞	12	0	37	8⅜
11	3⅛	35	4½	11	6⅛	36	1⅞	11	9⅛	36	11¼	12	0⅛	37	8¾
11	3¼	35	4⅞	11	6¼	36	2¼	11	9¼	36	11¾	12	0¼	37	9⅛
11	3⅜	35	5¼	11	6⅜	36	2⅝	11	9⅜	37	0⅛	12	0⅜	37	9½
11	3½	35	5⅝	11	6½	36	3	11	9½	37	0½	12	0½	37	9⅞
11	3⅝	35	6	11	6⅝	36	3½	11	9⅝	37	0⅞	12	0⅝	37	10¼
11	3¾	35	6⅜	11	6¾	36	3⅞	11	9¾	37	1¼	12	0¾	37	10⅝
11	3⅞	35	6¾	11	6⅞	36	4¼	11	9⅞	37	1⅝	12	0⅞	37	11⅛
11	4	35	7¼	11	7	36	4⅝	11	10	37	2	12	1	37	11½
11	4⅛	35	7⅝	11	7⅛	36	5	11	10⅛	37	2½	12	1⅛	37	11¾
11	4¼	35	8	11	7¼	36	5⅜	11	10¼	37	2⅞	12	1¼	38	0¼
11	4⅜	35	8⅜	11	7⅜	36	5⅞	11	10⅜	37	3¼	12	1⅜	38	0⅝
11	4½	35	8¾	11	7½	36	6¼	11	10½	37	3⅝	12	1½	38	1
11	4⅝	35	9⅛	11	7⅝	36	6⅝	11	10⅝	37	4	12	1⅝	38	1⅜
11	4¾	35	9½	11	7¾	36	7	11	10¾	37	4⅜	12	1¾	38	1⅞
11	4⅞	35	10	11	7⅞	36	7⅜	11	10⅞	37	4¾	12	1⅞	38	2¼
11	5	35	10⅜	11	8	36	7¾	11	11	37	5¼	12	2	38	2⅝
11	5⅛	35	10¾	11	8⅛	36	8⅛	11	11⅛	37	5⅝	12	2⅛	38	3
11	5¼	35	11⅛	11	8¼	36	8½	11	11¼	37	6	12	2¼	38	3⅜
11	5⅜	35	11½	11	8⅜	36	9	11	11⅜	37	6⅜	12	2⅜	38	3¾
11	5½	35	11⅞	11	8½	36	9⅜	11	11½	37	6¾	12	2½	38	4⅛
11	5⅝	36	0¼	11	8⅝	36	9¾	11	11⅝	37	7⅛	12	2⅝	38	4½

If a Hoop of larger diameter than 12 feet is required, double some number.

Observations on Tables relating to the Diameters and Circumferences of Circles.

I do not intend to enter into any labored argument to prove the general utility of these Tables, as their simplicity and clearness are sufficient to stamp their value to the artist and mechanic. It will be clearly perceived, on inspection, that the Table commences with as small a diameter as is generally used in hoops and rings, viz. one inch, and increases by the regular gradation of one-eighth of an inch, to upwards of twelve feet; and in the column marked Circumference, against each Diameter stand the respective circumferences: hence all that is necessary on inspecting these Tables is to enter into them with any proposed diameter or circumference, and an answer to the inquiry is immediately obtained.

Example.—Required the circumference of a circle, the diameter being 8 feet 7 7-8 inches?

In the column of circumferences, opposite the given diameter, stands 27 feet 2¼ inches, the circumference required.

But it will be necessary to observe, that in the formation of hoops and rings a contraction of the metal takes place. Now, the just allowance for this contraction is the exact thickness of the metal, which must be added to the diameter.

Ex.—In making a hoop whose diameter inside is 6 feet 9 1-8 inches, the thickness of the iron being ½ inch, this ½ inch must be added to the given diameter, which will make it 6 feet 9 5-8 inches; this will allow 1 5-8 inch for the contraction in bending in a hoop, of the above diameter, giving the circumference or length of iron required for the hoop, 21 feet 4 3-8 inches.

The foregoing example appertains to the formation of hoops or iron bent on the flat; but in the formation of rings or iron bent on the edge, the same rule must also be followed, only taking care to add the *breadth* instead of the thickness. As for example :

To make a ring whose inside diameter is 8 feet 2¼ inches, the *breadth* of the iron being 2¼ inches ; by adding the 2¼ inches to the given diameter, will increase it to 8 feet 4¾ inches ; opposite to this diameter in the column of circumferences stands 26 feet 4½ inches, being the length of iron necessary for the ring.

The foregoing observations relate more particularly to plain hoops and rings ; but as respects the hoops that are on the wheels of railway carriages, a difference must be observed, which is as follows : These hoops having a flange projecting on the one edge of the surface, it will be necessary, in addition to the thickness of the metal, to add two-thirds of the thickness of the flange to the diameter, as the flange side would contract considerably more than the plain surface ; this is supposing the tires are in a straight form, but, in general, they come from the iron-works in a curved state. In the latter case, it will be only necessary to add the thickness of the bare metal, as the aforesaid portion of the thickness of the flange is allowed for in the curve. It has been found that the curve may be exactly obtained, by using four times the circumference of the hoop as a radius.

If the tire has not been previously curved, it may easily be done in the operation of bending ; the smith must pay particular attention to this, or he will have his hoop bent in an angle.

But the practical utility of this Table is not confined to smiths alone ; to the millwright it will be found equally useful and expeditious, as on a bare inspection of the Table he may ascertain the diameter of any wheel that may be required to be made, the pitch and number of teeth being given.

Ex.—Suppose a wheel were ordered to be made to contain sixty teeth, the pitch of the teeth to be 3 7-8 inches, the dimensions of the wheel may be ascertained simply as follows :

Multiply the pitch of the tooth by the number of teeth the wheel is to contain, and the product will be the circumference of the wheel : thus

$$3\tfrac{7}{8} \text{ inches pitch of the tooth,}$$
$$10 \times 6 = 60 \text{ the number of teeth,}$$

$$\text{Feet} \quad 19 \quad 4\tfrac{1}{2} \quad \text{the circumference of the wheel.}$$

However, by inspecting the column marked Circumference, I find the nearest number to this is 19 feet 4 3-8 inches, which is the eighth of an inch less than the true circumference ; but if this 1-8 were divided into 60 equal parts, it would not make the difference of a single hair's-breadth in the size of each tooth ; so that it is sufficiently near for any practical purpose. The diameter answering to this circumference is 6 feet 2 inches ; consequently, with one-half of this number as a radius, the circumference of the wheel will be described.

The manner in which the foregoing Table of Circumferences is found is as follows : Taking the diameter at unity, we have by decimal proportion

$$\text{in.} \quad \text{in.}$$
$$\text{As } 1 : 3\text{·}1416 :: 1\text{·} : 3\text{·}1416,$$

and the decimal 1416 multiplied by 8, gives the circumference for 1 inch of diameter 3 1·8 inches.

In these Tables the number 3·1416 is divided by 8, which gives ·3927 This decimal proportion has been used as a constant, and the sum multiplied by 8 gives the excess above the decimal value in eighths of an inch

CIRCUMFERENCES FOR ANGLED IRON HOOPS.

ANGLE OUTSIDE.

Diam.		Circ.		Diam.		Circ.		Diam.		Circ.		Diam.		Circ.		Diam.		Circ.	
Ft	In.	Ft.	In.	Ft.	In.	Ft.	In.	Ft.	In.	Ft.	In.	Ft.	In.	Ft.	in.	Ft.	In.	Ft.	In.
	6	1	5½	1	6	4	4¾	2	6	7	3⅞	3	6	10	3	4	6	13	2¼
	¼	1	6¼		¼	4	5⅜		¼	7	4⅝		¼	10	3¾		¼	13	3
	½	1	7		½	4	6⅛		½	7	5⅜		½	10	4½		½	13	3¾
	¾	1	7¾		¾	4	6⅞		¾	7	6⅛		¾	10	5¼		¾	13	4⅜
	7	1	8½	1	7	4	7⅝	2	7	7	6⅞	3	7	10	6	4	7	13	5⅛
	¼	1	9¼		¼	4	8⅜		¼	7	7½		¼	10	6¾		¼	13	5⅞
	½	1	9⅞		½	4	9⅛		½	7	8¼		½	10	7½		½	13	6⅝
	¾	1	10⅝		¾	4	9⅞		¾	7	9		¾	10	8¼		¾	13	7⅜
	8	1	11⅜	1	8	4	10⅝	2	8	7	9¾	3	8	10	8⅞	4	8	13	8⅛
	¼	2	0⅛		¼	4	11⅜		¼	7	10½		¼	10	9⅝		¼	13	8⅞
	½	2	0⅞		½	5	0		½	7	11¼		½	10	10⅜		½	13	9⅝
	¾	2	1⅝		¾	5	0¾		¾	8	0		¾	10	11⅛		¾	13	10⅜
	9	2	2⅜	1	9	5	1½	2	9	8	0⅝	3	9	10	11⅞	4	9	13	11
	¼	2	3		¼	5	2¼		¼	8	1⅜		¼	11	0⅝		¼	13	11¾
	½	2	3¾		½	5	3		½	8	2⅛		½	11	1⅜		½	14	0½
	¾	2	4½		¾	5	3¾		¾	8	2⅞		¾	11	2		¾	14	1¼
	10	2	5¼	1	10	5	4⅜	2	10	8	3⅝	3	10	11	2¾	4	10	14	2
	¼	2	6		¼	5	5⅛		¼	8	4⅜		¼	11	3½		¼	14	2¾
	½	2	6¾		½	5	5⅞		½	8	5⅛		½	11	4¼		½	14	3⅜
	¾	2	7½		¾	5	6⅝		¾	8	5¾		¾	11	5		¾	14	4⅛
	11	2	8¼	1	11	5	7⅜	2	11	8	6½	3	11	11	5¾	4	11	14	4⅞
	¼	2	8⅞		¼	5	8⅛		¼	8	7¼		¼	11	6½		¼	14	5⅝
	½	2	9⅝		½	5	8⅞		½	8	8		½	11	7⅛		½	14	6⅜
	¾	2	10⅜		¾	5	9⅝		¾	8	8¾		¾	11	7⅞		¾	14	7⅛
1	0	2	11⅛	2	0	5	10¼	3	0	8	9½	4	0	11	8⅝	5	0	14	7⅞
	¼	2	11⅞		¼	5	11		¼	8	10¼		¼	11	9⅜		¼	14	8½
	½	3	0⅝		½	5	11¾		½	8	11		½	11	10⅛		½	14	9¼
	¾	3	1⅜		¾	6	0½		¾	8	11⅝		¾	11	10⅞		¾	14	10
1	1	3	2	2	1	6	1¼	3	1	9	0⅜	4	1	11	11⅝	5	1	14	10¾
	¼	3	2¾		¼	6	2		¼	9	1⅛		¼	12	0¼		¼	14	11½
	½	3	3½		½	6	2¾		½	9	1⅞		½	12	1		½	15	0¼
	¾	3	4¼		¾	6	3⅜		¾	9	2⅝		¾	12	1¾		¾	15	1
1	2	3	5	2	2	6	4⅛	3	2	9	3⅜	4	2	12	2½	5	2	15	1⅝
	¼	3	5¾		¼	6	4⅞		¼	9	4⅛		¼	12	3¼		¼	15	2⅜
	½	3	6½		½	6	5⅝		½	9	4¾		½	12	4		½	15	3⅛
	¾	3	7⅛		¾	6	6⅜		¾	9	5½		¾	12	4¾		¾	15	3⅞
1	3	3	7⅞	2	3	6	7⅛	3	3	9	6¼	4	3	12	5½	5	3	15	4⅝
	¼	3	8⅝		¼	6	7⅞		¼	9	7		¼	12	6⅛		¼	15	5⅜
	½	3	9⅜		½	6	8½		½	9	7¾		½	12	6⅞		½	15	6⅛
	¾	3	10⅛		¾	6	9¼		¾	9	8½		¾	12	7⅝		¾	15	6⅞
1	4	3	10⅞	2	4	6	10	3	4	9	9¼	4	4	12	8⅜	5	4	15	7½
	¼	3	11⅝		¼	6	10¾		¼	9	9⅞		¼	12	9⅛		¼	15	8¼
	½	4	0¼		½	6	11½		½	9	10⅝		½	12	9⅞		½	15	9
	¾	4	1		¾	7	0¼		¾	9	11⅜		¾	12	10½		¾	15	9¾
1	5	4	1¾	2	5	7	1	3	5	10	0⅛	4	5	12	11¼	5	5	15	10½
	¼	4	2½		¼	7	1¾		¼	10	0⅞		¼	13	0		¼	15	11¼
	½	4	3¼		½	7	2⅜		½	10	1⅝		½	13	0¾		½	16	0
	¾	4	4		¾	7	3⅛		¾	10	2⅜		¾	13	1½		¾	16	0⅝

CIRCUMFERENCES FOR ANGLED IRON HOOPS.

ANGLE INSIDE.

Diam.		Circ.		Diam.		Circ.		Diam.		Circ.		Diam.		Circ.		Diam.		Circ.	
Ft.	In.	Ft.	In	Ft.	In.	Ft.	In.	Ft.	in.	Ft.	In.	Ft.	In.	Ft.	In.	Ft.	In.	Ft.	In.
	6	1	8½	1	6	5	1⅝	2	6	8	6¾	3	6	11	11¾	4	6	15	4⅞
	¼	1	9⅜		¼	5	2½		¼	8	7½		¼	12	0⅝		¼	15	5¾
	½	1	10¼		½	5	3¼		½	8	8⅜		½	12	1½		½	15	6⅝
	¾	1	11⅛		¾	5	4⅛		¾	8	9¼		¾	12	2⅜		¾	15	7½
	7	1	11⅞	1	7	5	5	2	7	8	10⅛	3	7	12	3¼	4	7	15	8¼
	¼	2	0¾		¼	5	5⅞		¼	8	11		¼	12	4⅛		¼	15	9⅛
	½	2	1⅝		½	5	6¾		½	8	11⅞		½	12	4⅞		½	15	10
	¾	2	2½		¾	5	7½		¾	9	0⅝		¾	12	5¾		¾	15	10⅞
	8	2	3⅜	1	8	5	8½	2	8	9	1½	3	8	12	6⅝	4	8	15	11¾
	¼	2	4¼		¼	5	9¼		¼	9	2⅜		¼	12	7½		¼	16	0⅝
	½	2	5		½	5	10⅛		½	9	3¼		½	12	8⅜		½	16	1½
	¾	2	5⅞		¾	5	11		¾	9	4⅛		¾	12	9¼		¾	16	2¼
	9	2	6¾	1	9	5	11⅞	2	9	9	5	3	9	12	10	4	9	16	3⅛
	¼	2	7⅝		¼	6	0¾		¼	9	5⅞		¼	12	10⅞		¼	16	4
	½	2	8½		½	6	1⅝		½	9	6⅝		½	12	11¾		½	16	4⅞
	¾	2	9⅜		¾	6	2½		¾	9	7½		¾	13	0⅝		¾	16	5¾
	10	2	10¼	1	10	6	3¼	2	10	9	8⅜	3	10	13	1½	4	10	16	6⅝
	¼	2	11		¼	6	4⅛		¼	9	9¼		¼	13	2⅜		¼	16	7⅜
	½	2	11⅞		½	6	5		½	9	10⅛		½	13	3¼		½	16	8¼
	¾	3	0¾		¾	6	5⅞		¾	9	11		¾	13	4		¾	16	9⅛
	11	3	1⅝	1	11	6	6¾	2	11	9	11⅞	3	11	13	4⅞	4	11	16	10
	¼	3	2½		¼	6	7⅝		¼	10	0⅝		¼	13	5¾		¼	16	10⅞
	½	3	3⅜		½	6	8⅜		½	10	1½		½	13	6⅝		½	16	11¾
	¾	3	4¼		¾	6	9¼		¾	10	2⅜		¾	13	7½		¾	17	0⅝
1	0	3	5	2	0	6	10⅛	3	0	10	3¼	4	0	13	8⅜	5	0	17	1½
	¼	3	5⅞		¼	6	11		¼	10	4⅛		¼	13	9¼		¼	17	2¼
	½	3	6¾		½	6	11⅞		½	10	5		½	13	10		½	17	3⅛
	¾	3	7⅝		¾	7	0¾		¾	10	5⅞		¾	13	10⅞		¾	17	4
1	1	3	8½	2	1	7	1⅝	3	1	10	6⅝	4	1	13	11¾	5	1	17	4⅞
	¼	3	9⅜		¼	7	2⅜		¼	10	7½		¼	14	0⅝		¼	17	5¾
	½	3	10⅛		½	7	3¼		½	10	8⅜		½	14	1½		½	17	6½
	¾	3	11		¾	7	4⅛		¾	10	9¼		¾	14	2⅜		¾	17	7⅜
1	2	3	11⅞	2	2	7	5	3	2	10	10⅛	4	2	14	3⅛	5	2	17	8¼
	¼	4	0¾		¼	7	5⅞		¼	10	11		¼	14	4		¼	17	9⅛
	½	4	1⅝		½	7	6¾		½	10	11¾		½	14	4⅞		½	17	10
	¾	4	2½		¾	7	7½		¾	11	0⅝		¾	14	5¾		¾	17	10⅞
1	3	4	3⅜	2	3	7	8⅜	3	3	11	1½	4	3	14	6⅝	5	3	17	11¾
	¼	4	4⅛		¼	7	9¼		¼	11	2⅜		¼	14	7½		¼	18	0½
	½	4	5		½	7	10⅛		½	11	3¼		½	14	8⅜		½	18	1⅜
	¾	4	5⅞		¾	7	11		¾	11	4⅛		¾	14	9⅛		¾	18	2¼
1	4	4	6¾	2	4	7	11¾	3	4	11	5	4	4	14	10	5	4	18	3⅛
	¼	4	7⅝		¼	8	0¾		¼	11	5¾		¼	14	10⅞		¼	18	4
	½	4	8½		½	8	1½		½	11	6⅝		½	14	11¾		½	18	4⅞
	¾	4	9⅜		¾	8	2⅜		¾	11	7½		¾	15	0⅝		¾	18	5¾
1	5	4	10⅛	2	5	8	3¼	3	5	11	8⅜	4	5	15	1½	5	5	18	6½
	¼	4	11		¼	8	4⅛		¼	11	9¼		¼	15	2¼		¼	18	7⅜
	½	4	11⅞		½	8	5		½	11	10⅛		½	15	3⅛		½	18	8¼
	¾	5	0¾		¾	8	5⅞		¾	11	10⅞		¾	15	4		¾	18	9⅛

OBSERVATIONS ON TABLE CONTAINING THE CIRCUMFERENCES FOR ANGLED IRON HOOPS.—ANGLE OUTSIDE.

As this Table will be useful to those smiths who chiefly work angled iron, it will be necessary to remark, that the observation made on Tables relating to the Diameters and Circumferences of Circles, respecting adding the thickness of the iron to the diameter, must be attended to in this, with this difference,—the breadth of the angle must be added to the diameter.

Example.—Suppose a hoop is wanted to be made of 2½ inch angled iron, whose diameter inside must be 12 inches. Here the 2½ inches must be added to the 12 inches, which raises the number to 1 foot 2½ inches. Looking into the Table, I find the circumference, or length of iron requisite for the hoop, is 3 feet 6½ inches.

OBSERVATIONS ON TABLE CONTAINING THE CIRCUMFERENCES FOR ANGLED IRON HOOPS.—ANGLE INSIDE.

The observations respecting this Table are the reverse to those on the preceding one,—viz. the breadth of the angle must be taken from the diameter,—for this reason, that the diameter is taken from outside to outside of the ring.

Suppose a ring is to be made of angled iron, whose diameter outside is to be 12 inches, the breadth of the angle 2½ inches; then, by taking 2½ inches from 12 inches, we have left 9½ inches. Looking into the Table in the column of diameters, I find in the circumference column, opposite 9½ inches, 2 feet 8½ inches, which is the length of iron necessary for the ring.

It has been already observed, that between angled and plain iron a considerable difference exists with regard to the proportion of the circumference to the diameter: this is owing to the angle or flange on one side of the bar, and when the iron is formed into a hoop: it contracts more or less, as the angle or flange may be inside or outside of the hoop. From repeated experiments on this subject, I have ascertained that the proportions of the diameters to the circumferences are as follows:— For the angle inside as 1 : 3·4248, and for the angle outside the hoop, as 1 : 2·9312 :: Diam : Circ'f.

Problem.—To find the circumference of an ellipse, or an oval hoop or ring.

Rule.—Add the length of the two axes together, and multiply the sum by 1·5708 for the circumference; or as it may be used in the Table of Circumferences, take half the sum of the axes as a diameter, with the breadth of the iron added, and enter the Table of Circumferences where it will be found.

Ex.—Required the circumference of an elliptical hoop, whose axes are 18½ and 13 inches, the thickness of the iron being 2½ inches.

$$19\tfrac{1}{2} + 13 = 31\tfrac{1}{2} \div 2 = 15\tfrac{3}{4} + 2\tfrac{1}{2} = 18\tfrac{1}{4} \text{ inches the diameter.}$$

Entering into the Table of Diameter with 18¼ inches, the circumference will be found to be 4 feet 9¼ inches.

In constructing elliptical hoops of angled iron, with the angle outside, reference must be made to the Tables for hoops of angled iron; the operation will be similar to the above example. But in hoops where the angle is inside, the thickness of the iron must be taken from half the sum of the axes.

Note.—It must be observed, that in the examples given in the Observations on Table relating to the Diameters and Circumferences of Circles, and also on hoops formed of angled iron, that those circumferences are nothing more than the ends of the iron meeting together; therefore, every smith must allow for the thickening of the ends of the metal previous to scarving the same in order to weld it.

16

SHIP AND RAILROAD SPIKES.

NUMBER OF IRON SPIKES PER 100 POUNDS.

Manufactured by PHILIP C. PAGE, *Mass., and Sold by* PAGE, BRIGGS & BABBITT, *Boston.*

Ship Spikes or Hatch Nails 1-4 in. sq're.		Ship Spikes or Hatch Nails 5-16 in. sq.		Ship Spikes or Deck Nails 3-8 in. sq're.		Ship Spikes 7-16 inch square.		Ship Spikes 1-2 inch square.		Ship Spikes 9-16 inch square.		Ship Spikes 5-8 inch square.	
size in inc	No. 1 0 0 lbs.	size in inc.	No. 1 0 0 lbs.	size in inc.	No. 1 0 0 lbs.	size in inc.	No. 1 0 0 lbs.	size in inc.	No. 1 0 0 lbs.	size in inc.	No. 1 0 0 lbs.	size in inc.	No. 1 0 0 lbs.
3	1900	3	1000	4	540	5	340	6	220	8	140	10	80
3½	1580	3½	960	4½	500	5½	310	6½	200	9	120	15	60
4	1320	4	800	5	460	6	300	7	190	10	110	—	—
4½	1220	4½	600	5½	420	6½	280	7½	180	11	100	—	—
5	1020	5	680	6	400	7	260	8	170	—	—	—	—
—	—	6	520	6½	320	7½	240	8½	160	—	—	—	—
—	—	—	—	—	—	8	220	9	150	—	—	—	—
—	—	—	—	—	—	—	—	10	140	—	—	—	—

Rail Road Spikes 9-16ths square 5½ inches 160 per 100 pounds.
Rail Road Spikes 1-2 inch " 5½ " 200 per 100 pounds.

BURDEN'S PATENT SPIKES AND HORSE SHOES.

Manufactured at the Troy Iron and Nail Factory, Troy, New York.

Boat Spikes.		Ship Spikes.		Hook Head.		Horse Shoes.	
Size in inches.	No. in 100 lbs.	Size in inches.	No. in 100 lbs.	Size in inches.	No. in 100 lbs.	Size in inches.	No. in 100 lbs.
3	1750	4	800	4 ×$\frac{3}{8}$	555	1	84
3½	1468	4½	650	4½×7-16	414	2	75
4	1257	5	437	5 ×½	252	3	65
4½	920	5½	430	5½×½	241	4	56
5	720	6	420	5½×9-16	187	5	39
5½	630	6½	377	6 ×9-16	172	—	—
6	497	7	275	6 ×⅝	138	—	—
6½	478	7½	250	7 ×9-16	140	—	—
7	362	8	174	8 ×⅝	110	—	—
7½	337	8½	163	—	—	—	—
8	295	9	155	—	—	—	—
8½	290	10	115	—	—	—	—
9	210	—	—	—	—	—	—
10	198	—	—	—	—	—	—

COPPERS.—*Dimensions and Weight from 1 to 208 Gallons.*

Inches lag to brim.	Gallons.	Weight in pounds.	Inches lag to brim.	Gallons.	Weight in pounds.	Inches lag to brim.	Gallons.	Weight in pounds.
9¾	1	1½	24	15	22½	29½	29	43½
12¼	2	3	24½	16	24	30	30	45
14	3	4½	25	17	25½	32	36	54
15½	4	6	25½	18	27	34	43	64½
16½	5	7½	26	19	28½	35	48	72
17½	6	9	26½	20	30	36	53	79½
18½	7	10½	26¾	21	31½	37	58	87
19½	8	12	27	22	33	38	63	94½
20¼	9	13½	27¼	23	34½	39	67	100½
21	10	15	27½	24	36	40	71	106½
21½	11	16½	27¾	25	37½	45	104	156
22	12	18	28	26	39	50	146	219
22½	13	19½	28½	27	40½	55	208	312
23¼	14	21	29	28	42			

COPPER TUBING. — *Weight of the usual Thickness.*

When the inside diameter, is ¼ of an inch, 3 ozs. ; ⅜ do., 5 ozs. ; ½ do., 6 ozs. ; ⅝ do., 8 ozs. ; ¾ do., 10 ozs. per foot.

BRASS, COPPER, STEEL AND LEAD.—*Weight of a Foot.*

	BRASS.		COPPER.		STEEL.		LEAD.	
Diam'ter and Side of Sq're.	Weight of Round.	Weight of Square.	Weight of Round.	Weight of Square.	Weight of Round.	Weight of Square.	Weight of Round.	Weight of Square.
Inches.	Lbs.	Lbs.	Lbs.	Lbs.	Lbs.	Lbs.	Lbs.	Lbs.
¼	.17	.22	.19	.24	.17	.21		
⅜	.39	.50	.42	.54	.38	.48		
½	.70	.90	.75	.96	.67	.85		
⅝	1.10	1.40	1.17	1.50	1.04	1.33		
¾	1.59	2.02	1.69	2.16	1.50	1.91		
⅞	2.16	2.75	2.31	2.94	2.05	2.61		
1	2.83	3.60	3.02	3.84	2.67	3.40	3.87	4.93
1⅛	3.58	4.56	3.82	4.86	3.38	4.34	4.90	6.25
1¼	4.42	5.63	4.71	6.	4.18	5.32	6.06	7.71
1⅜	5.35	6.81	5.71	7.27	5.06	6.44	7.33	9 33
1½	6.36	8.10	6.79	8.65	6.02	7.67	8.72	11.11
1⅝	7.47	9.51	7.94	10.15	7.07	9.	10.24	13.04
1¾	8.66	11.03	9.21	11.77	8.20	10.14	11 87	15.12
1⅞	9.95	12.66	10.61	13.52	9.41	11.98	13.63	17.36
2	11.32	14.41	12.08	15.38	10.71	13.63	15.51	19.75
2⅛	12.78	16.27	13.64	17.36	12.05	15.80	17.51	22.29
2¼	14.32	18.24	15.29	19.47	13.51	17.20	19.63	25.
2⅜	15.96	20.32	17.03	21.69	15 05	19.17	21.80	27.80
2½	17.68	22.53	18.87	24.03	16.68	21.21	24.24	30.86
2⅝	19.50	24.83	20.81	26.50	18.39	23.41	26.72	34.02
2¾	21.40	27.25	22.84	29.08	20.18	25.70	29.33	37.34
2⅞	23.39	29.78	24.92	31.79	22.06	28.10	32.05	40.81
3	25.47	32.43	27.18	34.61	24.23	30.60	34.90	44.44

CAST IRON.

Weight of a Foot in Length of Flat Cast Iron.

Width of Iron.	Thick, 1-4th inch.	Thick, 3-8ths inch	Thick, 1-2 inch.	Thick, 5-8ths inch.	Thick, 3-4ths inch.	Thick, 7-8ths inch.	Thick, 1 inch.
Inches.	Pounds.	Pounds.	Pounds.	Pounds.	Pounds.	Pounds.	Pounds.
2	1·56	2·34	3·12	3·90	4·68	5·46	6·25
2¼	1·75	2·63	3·51	4·39	5·27	6·15	7·03
2½	1·95	2·92	3·90	4·88	5·85	6·83	7·81
2¾	2·14	3·22	4·29	5·37	6·44	7·51	8·59
3	2·34	3·51	4·68	5·85	7·03	8·20	9·37
3¼	2·53	3·80	5·07	6·34	7·61	8·88	10·15
3½	2·73	4·10	5·46	6·83	8·20	9·57	10·93
3¾	2·93	4·39	5·85	7·32	8·78	10·25	11·71
4	3·12	4·68	6·25	7·81	9·37	10·93	12·50
4¼	3·32	4·97	6·64	8·30	9·96	11·62	13·28
4½	3·51	5 27	7·03	8·78	10 54	12·30	14·06
4¾	3·71	5·56	7·42	9·27	11·13	12·98	14·84
5	3·90	5·86	7·81	9·76	11·71	13·67	15·62
5¼	4·10	6·15	8·20	10·25	12·30	14·35	16·40
5½	4·29	6·44	8·59	10·74	12·89	15·03	17·18
5¾	4·49	6·73	8·98	11·23	13·46	15·72	17·96
6	4·68	7·03	9·37	11·71	14·06	16·40	18·75

CAST IRON.

Weight of a Superficial Foot from ¼ to 2 inches thick.

Size.	Weight.	Size.	Weight.	Size.	Weight.	Size.	Weight.	Size	Weight.
Ins.	Pounds.	Ins.	Pounds.	Ins.	Pounds.	Ins.	Pounds.	Ins.	Pounds.
¼	9.37	⅝	23.43	1	37.50	1⅜	51.56	1¾	65.62
⅜	14.06	¾	28.12	1⅛	42.18	1½	56.25	1⅞	70.31
½	18.75	⅞	32.81	1¼	46.87	1⅝	60.93	2	75

CAST IRON, COPPER, BRASS, AND LEAD BALLS.

Weight of Cast Iron, Copper, Brass, and Lead Balls, from 1 inch to 12 inches in Diameter.

Diam.	Cast Iron.	Copper.	Brass.	Lead.	Diam.	Cast Iron.	Copper.	Brass.	Lead.
Ins.	pounds.	pounds.	pounds.	pounds.	Inches.	pounds.	pounds.	pounds.	pounds.
1	·136	·166	·158	·214	7	46·76	57·1	54·5	73·7
1½	·46	·562	·537	·727	7½	57·52	70·0	67·11	90·0
2	1·09	1·3	1·25	1·7	8	69·81	85·2	81·4	110·1
2½	2·13	2·60	2·50	3·35	8½	83·73	102·3	100·0	132·3
3	3·68	4·5	4·3	5·8	9	99.4	121·3	115·9	156·7
3½	5·84	7·14	6·82	9·23	9½	116·9	143·0	136·4	184·7
4	8·72	10·7	10·2	13·8	10	136·35	166·4	159·0	215·0
4½	12·42	15·25	14·5	19·6	10½	157·84	193·0	184·0	2[illegible]·0
5	17·04	20·8	19·9	26·9	11	181·48	221·8	211·8	286·7
5½	22·68	27·74	26·47	36·0	11½	207·37	253·5	242·0	327·7
6	29·45	35·9	34·3	46·4	12	235·62	288·1	275·0	372:3
6½	37·44	45·76	43·67	59·13					

CAST IRON. — *Weight of a Foot in Length of Square and Round.*

SQUARE.				ROUND.			
Size.	Weight.	Size.	Weight.	Size.	Weight.	Size.	Weight.
Inches Square	Pounds.	Inches Square.	Pounds.	Inches Diam.	Pounds.	Inches Diam.	Pounds.
½	·78	4⅞	74·26	½	·61	4⅞	58·32
⅝	1·22	5	78·12	⅝	·95	5	61·35
¾	1·75	5⅛	82·08	¾	1·38	5⅛	64·46
⅞	2·39	5¼	86·13	⅞	1 87	5¼	67·64
1	3·12	5⅜	90·28	1	2·45	5⅜	70·09
1⅛	3·95	5½	94·53	1⅛	3 10	5½	74·24
1¼	4·88	5⅝	98·87	1¼	3 83	5⅝	77·65
1⅜	5·90	5¾	103·32	1⅜	4·64	5¾	81·14
1½	7·03	5⅞	107·86	1½	5·52	5⅞	84·71
1⅝	8·25	6	112·50	1⅝	6·48	6	88·35
1¾	9·57	6¼	122·08	1¾	7·51	6¼	95·87
1⅞	10·98	6½	132·03	1⅞	8·62	6½	103·69
2	12 50	6¾	142·38	2	9·81	6¾	111 82
2⅛	14·11	7	153·12	2⅛	11·08	7	120·26
2¼	15·81	7¼	164·25	2¼	12·42	7¼	129·
2⅜	17·62	7½	175·78	2⅜	13·84	7½	138·05
2½	19·53	7¾	187·68	2½	15·33	7¾	147·41
2⅝	21·53	8	200·	2⅝	16·91	8	157·08
2¾	23·63	8¼	212·56	2¾	18·56	8¼	167·05
2⅞	25·83	8½	225·78	2⅞	20·28	8½	177·10
3	28·12	8¾	239·25	3	22·08	8¾	187·91
3⅛	30·51	9	253·12	3⅛	23·96	9	198·79
3¼	33·	9¼	267·38	3¼	25·92	9¼	210·
3⅜	35·59	9½	282·	3⅜	27·95	9½	221·50
3½	38·28	9¾	297·07	3½	30·06	9¾	233·31
3⅝	41·06	10	312·50	3⅝	32·25	10	245·43
3¾	43·94	10¼	328·32	3¾	34·51	10¼	257·86
3⅞	46·92	10½	344·53	3⅞	36·85	10½	270·59
4	50·	10¾	361·13	4	39·27	10¾	283·63
4⅛	53·14	11	378·12	4⅛	41·76	11	296·97
4¼	56·44	11¼	395·50	4¼	44·27	11¼	310·63
4⅜	59·81	11½	413·28	4⅜	46·97	11½	324·59
4½	63·28	11¾	431·44	4½	49·70	11¾	338·85
4⅝	66·84	12	450·	4⅝	52·50	12	353·43
4¾	70·50			4¾	55·37		

STEEL. — *Weight of a Foot in Length of Flat.*

Size.	Thick, 1-4 inch.	Thick, 3-8ths.	Thick, 1-2 inch.	Thick, 5-8ths.	Size.	Thick, 1-4 inch.	Thick, 3-8ths.	Thick, 1-2 inch.	Thick, 5-8ths.
Inches	pounds.	pounds.	pounds.	pounds.	Inches.	pounds.	pounds.	pounds.	pounds.
1	·852	1·27	1·70	2.13	2½	2·13	3·20	4·26	5·32
1⅛	·958	1·43	1·91	2·39	2¾	2·34	3·51	4·68	5·85
1¼	1·06	1·59	2·13	2·66	3	2·55	3·83	5·11	6·39
1⅜	1 17	1·75	2·34	2·92	3¼	2·77	4·15	5·53	6·92
1½	1·27	1·91	2·55	3·19	3½	2·98	4·47	5·98	7·45
1¾	1·49	2·23	2·98	3·72	3¾	3·19	4·79	6·38	7·98
2	1·70	2·55	3·40	4·26	4	3·40	5·10	6·80	8·52
2¼	1·91	2·87	3·83	4·79					

WEIGHTS OF ROLLED IRON

Per lineal foot, in pounds and decimal parts, of sections of Parallel Angle, Taper Angle, Parallel T, *Taper* T, *and Sash Iron and Rails.*

Table I. — Parallel Angle Iron, of Equal Sides.

Length of sides. A B, in inches.	Uniform thickness throughout.	Weight of one lineal foot.
in.	*in.*	
3	⅜	8·0
2¾	⅜	7 0
2½	⅜	5·75
2¼	5-16ths	4·5
2	¼ full	3·75
1¾	¼	3·0
1½	¼	2 5
1⅜	No. 6 wire guage	1·75
1¼	8	1·5
1⅛	9	1·25
1	10	1·0
⅞	10	·875
¾	11	·625
⅝	11	·563
½	12	·5

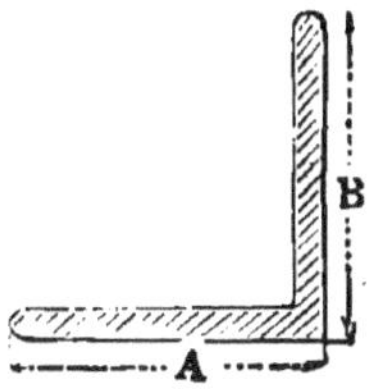

Table II.— Parallel Angle Iron, of Unequal Sides.

L'gth of side A in inches.	L'gth of side B in inches.	Uniform thickness throughout.	Weight of 1 lineal foot.
in.	*in.*	*in.*	
3½	5	⅜	9·75
3	5	⅜	8·75
3	4	5-16ths	7·5
2¼	4	5-16ths	6·75
2¼	4	¼	5·75
2	4	¼	5·5
2½	3	¼	4·75
2	2½	¼	3·375
1½	2	¼	2·875
1½	2	3-16ths	2·25

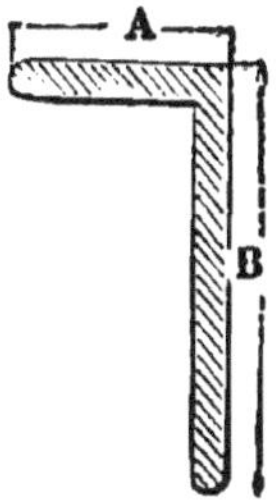

Table III.— Taper Angle Iron, of Equal Sides

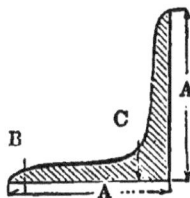

L'gth of sides A A, in inches.	Thickness of edges at B.	Thickness of root at C.	Weight of 1 lineal foot.
in.	*in.*	*in.*	
4	½	⅝	14·0
3	½	⅝	10·375
2¾	7-16ths	9-16ths	8·25
2½	⅜	½	6·5
2¼	5-16ths, full	7-16ths	5·0
2	¼ full	5-16ths full	3·875
1¾	¼	5-16ths	3·25
1½	¼ bare	5-16th, bare	2 625

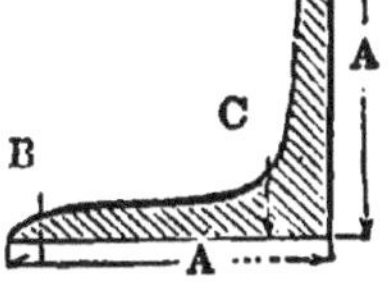

WEIGHTS OF PARALLEL AND TAPER T IRON.

Table IV.—PARALLEL T IRON, OF UNEQUAL WIDTH AND DEPTH.

Width of top table A.	Total depth B.	Uniform thickness top table C	Uniform thickness of rib D.	Weight of one lineal foot.
in.	*in.*	*in.*	*in.*	
5	6	½	½	15·75
4½	3¼	½	9-16ths	13·25
4	3	⅜	⅜	8·875
3½	3	⅜	⅜	8·25
3½	4	½	½	12 5
2½	3	⅜	⅜	7·0
2¼	2	5-16ths	⅜ full	4·5
2	1½	5-16ths	5-16ths	4·0
1¾	2	¼	¼	3·125
1½	2	¼	¼	2·875
1¼	1½	¼	¼	2·375
1	1¼	3-16ths	3-16ths	1·5
¾	1	3-16ths	3-16ths	1·125

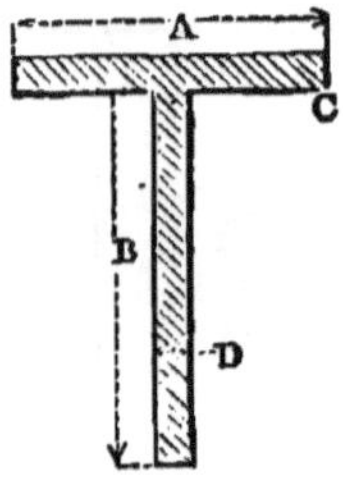

Table V.—PARALLEL T IRON, OF EQUAL DEPTH AND WIDTH.

Width of top table, and total depth A, A.	Uniform thickness throughout	Weight of one lineal foot.
in.	*in.*	
6	½	
5	7-16ths	13·75
4	⅜	9·75
3½	⅜	8·5
3	⅜	7·5
2½	5-16ths	4·625
2¼	5-16ths	4·5
2	5-16ths	3·75
1¾	¼	3·0
1½	¼	2·25
1¼	¼	1·75
1	3-16ths	1·0
⅞	⅛	·725
¾	⅛	·625

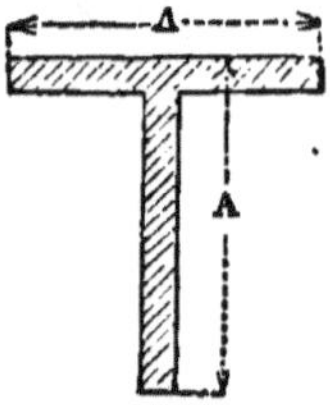

Table VI.—TAPER T IRON

Width of top table A	Total depth B.	Thickness of top table at root C.	Thickness of top table at edges D.	Uniform thickness of rib E.	Weight of one lin. foot.
in.	*in.*	*in.*	*in.*	*in.*	
3	3¼	½	⅜	7-16ths	8·0
3	2⅝	7-16ths	⅜	½	8·0
2	3	7-16ths	5-16ths	5-16ths	5·25
2½	2½	⅝	½	½ full	6·5
2	1½	⅜ full	5-16ths	⅜	3·5
2	1½	5-16ths	¼	¼	2·875

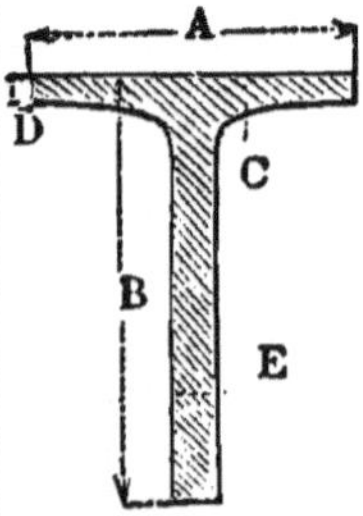

WEIGHT OF SASHES AND RAILS.

Table VII. — Sash Iron.

Total depth A.	Depth of rebate B.	Width at edge C.	greatest width D.	Weight of one lineal foot.
in.	*in.*		*in.*	
2	1	No. 9 w. guage	5-8ths	1·75
1¾	¾	7	9-16ths	1·625
1½	¾	6	9-16ths	1·25
1⅜	⅝	10	9-16ths	1·125
1¼	⅝	10	9-16ths	1·0
1	½	⅛	½	·75

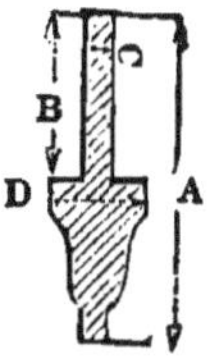

Table VIII — Rails equal top and bottom Tables.

Depth A in inches.	Width across top and bottom, BB, in inches.	Thickness of rib C.	Weight of 1 lin. foot
in.	*in.*	*in.*	
5	2⅝	¾	25·0
4½	2½	¾	23·33
4½	2½	⅝	21·66

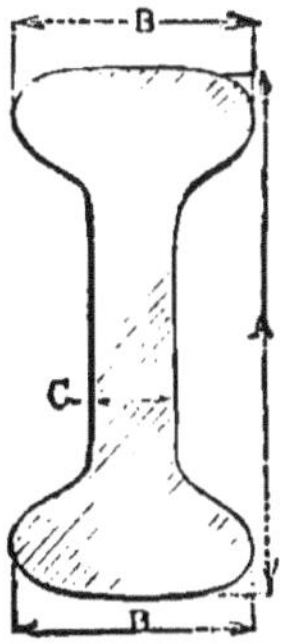

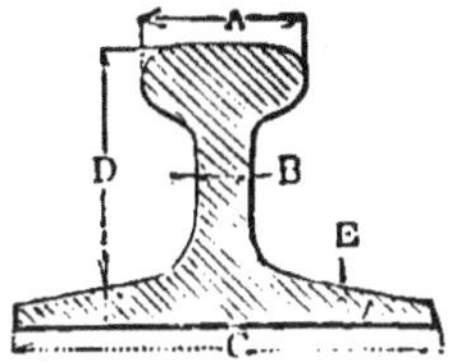

Table IX. — Temporary Rails.

Top width A.	Rib width B.	Bed width C.	Total depth D.	Thickness of bed E.	Weight of 1 lin. foot
in.	*in.*	*in.*	*in.*	*in.*	
1½	⅝	3	2	7-16ths	9·0
1¾	⅝	3	2½	½	12·0
1⅞	⅝	4	3	½	16·0
2	⅝	4	3	½	17·33

WEIGHT OF A LINEAL FOOT OF MALLEABLE RECTANGULAR OR FLAT IRON.

From an Eighth of an Inch to Three Inches Thick.

T designates the thickness, B. the breadth.

T.	B.	Weight lbs.	ozs.	T.	B.	Weight lbs.	ozs.	T.	B.	Weight lbs.	ozs.	T.	B.	Weight lbs.	ozs.
in.	in	lbs.	ozs.	in.	in.	lbs.	ozs.	in.	in.	lbs.	ozs.	in.	in.	lbs.	ozs.
1/8	1/4	0	1.6	1/8	10 3/4	4	7.3	1/4	9 1/2	7	14.1	3/8	8 3/4	10	13.8
	3/8	0	2.4		11	4	9.0		9 3/4	8	1.4		9	11	2.8
	1/2	0	3.3		11 1/4	4	10.7		10	8	4.8		9 1/4	11	7.8
	5/8	0	4.1		11 1/2	4	12.3		10 1/4	8	8.1		9 1/2	11	12.7
	3/4	0	5.0		11 3/4	4	14.0		10 1/2	8	11.4		9 3/4	12	1.7
	7/8	0	5.8		12	4	15.6		10 3/4	8	14.7		10	12	6.7
	1	0	6.6	—	—	—	—		11	9	2.0		10 1/4	12	11.6
	1 1/4	0	8.3	1/4	1/2	0	6.6		11 1/4	9	5.4		10 1/2	13	0.6
	1 1/2	0	9.9		5/8	0	8.3		11 1/2	9	8.7		10 3/4	13	5.6
	1 3/4	0	11.6		3/4	0	10.0		11 3/4	9	12.0		11	13	10.5
	2	0	13.2		7/8	0	11.6		12	9	15.3		11 1/4	13	15.5
	2 1/4	0	14.9	1	1	0	13.2						11 1/2	14	4.5
	2 1/2	1	0.6		1 1/4	1	0.6	3/8	3/4	0	14.9		11 3/4	14	9.4
	2 3/4	1	2.2		1 1/2	1	3.9		7/8	1	1.3		12	14	14.4
	3	1	3.9		1 3/4	1	7.2	1	1	1	3.8				
	3 1/4	1	5.5		2	1	10.5		1 1/4	1	8.8	1/2	1	1	10.4
	3 1/2	1	7.2		2 1/4	1	13.8		1 1/2	1	13.8		1 1/4	2	1.1
	3 3/4	1	8.9		2 1/2	2	1.2		1 3/4	2	2.7		1 1/2	2	7.7
	4	1	10.5		2 3/4	2	4.5		2	2	7.7		1 3/4	2	14.3
	4 1/4	1	12.2		3	2	7.8		2 1/4	2	12.7		2	3	4.9
	4 1/2	1	13.8		3 1/4	2	11.1		2 1/2	3	1.6		2 1/4	3	11.6
	4 3/4	1	15.5		3 1/2	2	14.4		2 3/4	3	6.6		2 1/2	4	2.2
	5	2	1.2		3 3/4	3	1.8		3	3	11.6		2 3/4	4	8.8
	5 1/4	2	2.8		4	3	5.1		3 1/4	4	0.5		3	4	15.4
	5 1/2	2	4.5		4 1/4	3	8.4		3 1/2	4	5.5		3 1/4	5	6.1
	5 3/4	2	6.1		4 1/2	3	11.7		3 3/4	4	10.5		3 1/2	5	12.7
	6	2	7.8		4 3/4	3	15.0		4	4	15.4		3 3/4	6	3.3
	6 1/4	2	9.5		5	4	2.4		4 1/4	5	4.4		4	6	9.9
	6 1/2	2	11.1		5 1/4	4	5.7		4 1/2	5	9.4		4 1/4	7	0.6
	6 3/4	2	12.8		5 1/2	4	9.0		4 3/4	5	14.3		4 1/2	7	7.2
	7	2	14.4		5 3/4	4	12.3		5	6	3.3		4 3/4	7	13.8
	7 1/4	3	0.1		6	4	15.6		5 1/4	6	8.3		5	8	4.4
	7 1/2	3	1.8		6 1/4	5	3.0		5 1/2	6	13.2		5 1/4	8	11.1
	7 3/4	3	3.4		6 1/2	5	6.3		5 3/4	7	2.2		5 1/2	9	1.7
	8	3	5.1		6 3/4	5	9.6		6	7	7.2		5 3/4	9	8.3
	8 1/4	3	6.7		7	5	13.0		6 1/4	7	12.2		6	9	14.9
	8 1/2	3	8.4		7 1/4	6	0.2		6 1/2	8	1.1		6 1/4	10	5.6
	8 3/4	3	10.1		7 1/2	6	3.6		6 3/4	8	6.1		6 1/2	10	12.2
	9	3	11.7		7 3/4	6	7.0		7	8	11.1		6 3/4	11	2.8
	9 1/4	3	13.4		8	6	10.2		7 1/4	9	0.0		7	11	9.4
	9 1/2	3	15.0		8 1/4	6	13.5		7 1/2	9	5.0		7 1/4	12	0.0
	9 3/4	4	7		8 1/2	7	0.8		7 3/4	9	10.0		7 1/2	12	6.7
	10	4	2.4		8 3/4	7	4.2		8	9	14.9		7 3/4	12	13.3
	10 1/4	4	4.0		9	7	7.5		8 1/4	10	3.9		8	13	3.9
	10 1/2	4	5.7		9 1/4	7	10.8		8 1/2	10	8.9		8 1/4	13	10.5

T. designates the thickness. B. the breadth.

T.	B.	Weight.		T.	B.	Weight.		T.	B.	Weight.		T.	B.	Weight.	
in.	in.	lbs.	ozs.	in.	in.	lbs.	ozs.	in.	in.	lbs.	ozs.	in.	in.	lbs.	ozs.
½	8½	14	1·2	⅝	9½	19	10·6	¾	10¾	26	11·2	1	2	6	10·0
	8¾	14	7·8		9¾	20	2.9		11	27	5·1		2¼	7	7·2
	9	14	14·4		10	20	11·2		11¼	27	15·1		2½	8	4.4
	9¼	15	5·0		10¼	21	3·4		11½	28	9·0		2¾	9	1·7
	9½	15	11·7		10½	21	11·7		11¾	29	3·0		3	9	14·7
	9¾	16	2·3		10¾	22	4·0		12	29	12·9		3¼	10	12·2
	10	16	8·9		11	22	12·3						3½	11	9·4
	10¼	16	15·5		11¼	23	4·6	⅞	1¾	5	1·1		3¾	12	6·7
	10½	17	6·2		11½	23	12·8		2	5	12·7		4	13	3·9
	10¾	17	12·8		11¾	24	5·1		2¼	6	8·3		4¼	14	1·2
	11	18	3·4		12	24	13·4		2½	7	3.9		4½	14	14·4
	11¼	18	10·0						2¾	7	15·5		4¾	15	11·7
	11½	19	0·7	¾	1½	3	11·6		3	8	11.1		5	16	8·9
	11¾	19	7·3		1¾	4	5·5		3¼	9	6·7		5¼	17	6·2
	12	19	13·9		2	4	15·4		3½	10	2.2		5½	18	3·4
					2¼	5	9·4		3¾	10	13·8		5¾	19	0·7
⅝	1¼	2	9·4		2½	6	3·3		4	11	9·4		6	19	13·9
	1½	3	.1·6		2¾	6	13·2		4¼	12	5·0		6¼	20	11·2
	1¾	3	9·9		3	7	7·2		4½	13	0·6		6½	21	8·4
	2	4	2·2		3¼	8	1·1		4¾	13	12·2		6¾	22	5.7
	2¼	4	10·5		3½	8	11·1		5	14	7 8		7	23	2·9
	2½	5	2·8		3¾	9	5·0		5¼	15	3·4		7¼	24	0·2
	2¾	5	11·0		4	9	14·9		5½	15	15·0		7½	24	13·4
	3	6	3.3		4¼	10	8·9		5¾	16	10·6		7¾	25	10·6
	3¼	6	11·6		4½	11	2·8		6	17	6·2		8	26	7·9
	3½	7	3·9		4¾	11	12·7		6¼	18	1·8		8¼	27	5·1
	3¾	7	12 2		5	12	6 7		6½	18	13 4		8½	28	2·4
	4	8	4·4		5¼	13	0·6		6¾	19	8·9		8¾	28	15·6
	4¼	8	12.7		5½	13	10·6		7	20	4·5		9	29	12·9
	4½	9	5·0		5¾	14	4·5		7¼	21	0·1		9¼	30	10·1
	4¾	9	13·3		6	14	14·4		7½	21	11.7		9½	31	7·4
	5	10	5·6		6¼	15	8·4		7¾	22	7·3		9¾	32	4·6
	5¼	10	13·8		6½	16	2·3		8	23	2.9		10	33	1·9
	5½	11	6·1		6¾	16	12·2		8¼	23	14·5		10¼	33	15·1
	5¾	11	14·4		7	17	6 2		8½	24	10·1		10½	34	12·4
	6	12	6·7		7¼	18	0·1		8¾	25	5·7		10¾	35	9.6
	6¼	12	15·0		7½	18	10·0		9	26	1·3		11	36	6 9
	6½	13	7·2		7¾	19	4·0		9¼	26	12·9		11¼	37	4·1
	6¾	13	15·5		8	19	13·9		9½	27	8.5		11½	38	1·4
	7	14	7·8		8¼	20	7·8		9¾	28	4·0		11¾	38	14·6
	7¼	15	0.1		8½	21	1·8		10	28	15·6		12	39	11·9
	7½	15	8·4		8¾	21	11·7		10¼	29	11·2				
	7¾	16	0·6		9	22	5.7		10½	30	6·8	1⅛	2¼	8	6·1
	8	16	8·9		9¼	22	15·6		10¾	31	2·4		2½	9	5·0
	8¼	17	1·2		9½	23	9·5		11	31	14 0		2¾	10	3·9
	8½	17	9·5		9¾	24	3.5		11¼	32	9·6		3	11	2·8
	8¾	18	1·8		10	24	13·4		11½	33	5·2		3¼	12	1·7
	9	18	10.0		10¼	25	7·3		11¾	34	0·8		3½	13	0·6
	9¼	19	2 3		10½	26	1·3		12	34	12.4		3¾	13	15·5

T. designates the thickness, B. the breadth.

T.	B.	Weight		T.	B.	Weight		T.	B.	Weight		T.	B.	Weight	
in.	in.	lbs.	ozs.	in.	in.	lbs.	ozs.	in	in.	lbs.	ozs.	in.	in.	lbs.	ozs.
1⅛	4	14	14·4	1¼	6¼	25	14·0	1⅜	8¾	39	13·5	1½	11½	57	2·1
	4¼	15	13·3		6½	26	14·5		9	40	15·7		11¾	58	5·9
	4½	16	12·2		6¾	27	15·1		9¼	42	2·0		12	59	9·8
	4¾	17	11·1		7	28	15·6		9½	43	4·2				
	5	18	10·0		7¼	30	0·2		9¾	44	6·4	1⅝	3¼	17	7·8
	5¼	19	8·9		7½	31	0·8		10	45	8·6		3½	18	13·4
	5½	20	7·8		7¾	32	1·3		10¼	46	10·8		3¾	20	2·9
	5¾	21	6·8		8	33	1·9		10½	47	13·0		4	21	8·4
	6	22	5·7		8¼	34	2·4		10¾	48	15·2		4¼	22	13·9
	6¼	23	4·6		8½	35	3·0		11	50	1·5		4½	24	3·5
	6½	24	3·5		8¾	36	3·6		11¼	51	3·7		4¾	25	9·0
	6¾	25	2·4		9	37	4·1		11½	52	5·9		5	26	14·5
	7	26	1·3		9¼	38	4·7		11¾	53	8·1		5¼	28	4·0
	7¼	27	0·2		9½	39	5·2		12	54	10·3		5½	29	9·6
	7½	27	15·1		9¾	40	5·8	1½	3	14	14·4		5¾	30	15·1
	7¾	28	14·0		10	41	6·4		3¼	16	2·3		6	32	4·6
	8	29	12·9		10¼	42	6·9		3½	17	6·2		6¼	33	10·2
	8¼	30	11·8		10½	43	7·5		3¾	18	10·0		6½	34	15·7
	8½	31	10·7		10¾	44	8·0		4	19	13·9		6¾	36	5·2
	8¾	32	9·6		11	45	8·6		4¼	21	1·8		7	37	10·7
	9	33	8·5		11¼	46	9·2		4½	22	5·7		7¼	39	0·3
	9¼	34	7·4		11½	47	9·7		4¾	23	9·5		7½	40	5·8
	9½	35	6·3		11¾	48	10·3		5	24	13·4		7¾	41	11·3
	9¾	36	5·2		12	49	10·8		5¼	26	1·3		8	43	0·9
	10	37	4·1		—				5½	27	5·1		8¼	44	6·4
	10¼	38	3·0	1⅜	2¾	12	8·3		5¾	28	9·0		8½	45	11·9
	10½	39	1·9		3	13	10·6		6	29	12·9		8¾	47	1·4
	10¾	40	0·8		3¼	14	12·8		6¼	31	0·8		9	48	7·0
	11	40	15·7		3½	15	15·0		6½	32	4·6		9¼	49	12·5
	11¼	41	14·6		3¾	17	1·2		6¾	33	8·5		9½	51	2·0
	11½	42	13·5		4	18	3·4		7	34	12·4		9¾	52	7·6
	11¾	43	12·4		4¼	19	5·6		7¼	36	0·2		10	53	13·1
	12	44	11·4		4½	20	7·8		7½	37	4·1		10¼	55	2·6
—					4¾	21	10·1		7¾	38	8·0		10½	56	8·1
1¼	2½	10	5·6		5	22	12·3		8	39	11·9		10¾	57	13·7
	2¾	11	6·1		5¼	23	14·5		8¼	40	15·7		11	59	3·2
	3	12	6·7		5½	25	0·7		8½	42	3·6		11¼	60	8·7
	3¼	13	7·2		5¾	26	2·9		8¾	43	7·5		11½	61	14·2
	3½	14	7·8		6	27	5·1		9	44	11·4		11¾	63	3·8
	3¾	15	8·4		6¼	28	7·4		9¼	45	15·2		12	64	9·3
	4	16	8·9		6½	29	9·6		9½	47	3·1	—			
	4¼	17	9·5		6¾	30	11·8		9¾	48	7·0	1¾	3¼	20	4·5
	4½	18	10·0		7	31	14·0		10	49	10·8		3½	21	11·7
	4¾	19	10·6		7¼	33	0·2		10¼	50	14·7		4	23	2·9
	5	20	11·2		7½	34	2·4		10½	52	2·6		4¼	24	10·1
	5¼	21	11·7		7¾	35	4·7		10¾	53	6·5		4½	26	1·3
	5½	22	12·3		8	36	6·9		11	54	10·3		4¾	27	8·5
	5¾	23	12·8		8¼	37	9·1		11¼	55	14 2		5	28	15·6
	6	24	13·4		8½	38	11·3		—				5¼	30	6·8

T. designates the thickness, B. the breadth.

T.	B.	Weight.		T.	B.	Weight.		T.	B.	Weight.		T.	B.	Weight.	
in.	in.	lbs.	ozs.	in.	in.	lbs.	ozs.	in.	in.	lbs.	ozs.	in.	in.	lbs.	ozs.
1¾	5½	31	14·0	1⅞	9	55	14·2	2⅛	4½	31	10·7	2¼	8¾	65	3·2
	5¾	33	5·2		9¼	57	7·0		4¾	33	6·8		9	67	1·0
	6	34	12·4		9½	58	15·9		5	35	3·0		9¼	68	14·9
	6¼	36	3·6		9¾	60	8·7		5¼	36	15·2		9½	70	12·7
	6½	37	10·7		10	62	1·6		5½	38	11·3		9¾	72	10·5
	6¾	39	1·9		10¼	63	10·4		5¾	40	7·5		10	74	8·3
	7	40	9·1		10½	65	3·2		6	42	3·6		10¼	76	6·1
	7¼	42	0·3		10¾	66	12·1		6¼	43	15·8		10½	78	3·9
	7½	43	7·5		11	68	4·9		6½	45	11·9		10¾	80	1·7
	7¾	44	14.7		11¼	69	13·8		6¾	47	8·1		11	81	15·5
	8	46	5·8		11½	71	6·6		7	49	4·2		11¼	83	13·3
	8¼	47	13·0		11¾	72	15·4		7¼	51	0·4		11½	85	11·1
	8½	49	4·2		12	74	8·3		7½	52	12·5		11¾	87	8·9
	8¾	50	11·4						7¾	54	8·7		12	89	6·7
	9	52	2·6	2	4	26	7·9		8	56	4·8				
	9¼	53	9·8		4¼	28	2·4		8¼	58	1·0	2⅜	4¾	37	5·8
	9½	55	1·0		4½	29	12·9		8½	59	13·1		5	39	5·2
	9¾	56	8·1		4¾	31	7·4		8¾	61	9·3		5¼	41	4·7
	10	57	15·3		5	33	1·9		9	63	5·4		5½	43	4·2
	10¼	59	6·5		5¼	34	12·4		9¼	65	1·6		5¾	45	3·6
	10½	60	13·7		5½	36	6·9		9½	66	13·7		6	47	3·1
	10¾	62	4·9		5¾	38	1·4		9¾	68	9·9		6¼	49	2·6
	11	63	12·1		6	39	11·9		10	70	6·0		6½	51	2·0
	11¼	65	3·2		6¼	41	6·4		10¼	72	2·2		6¾	53	1·5
	11½	66	10·4		6½	43	0·9		10½	73	14·3		7	55	1·0
	11¾	68	1·6		6¾	44	11·4		10¾	75	10·5		7¼	57	0·4
	12	69	8·8		7	46	5·8		11	77	6·6		7½	58	15·9
					7¼	48	0·3		11¼	79	2·8		7¾	60	15·3
1⅞	3¾	23	4·6		7½	49	10·8		11½	80	15·0		8	62	14·8
	4	24	13·4		7¾	51	5·3		11¾	82	11·1		8¼	64	14·3
	4¼	26	6·2		8	52	15·8		12	84	7·3		8½	66	13·7
	4½	27	15.1		8¼	54	10·3						8¾	68	13·2
	4¾	29	7·9		8½	56	4·8	2¼	4½	33	8·5		9	70	12·7
	5	31	0·8		8¾	57	15·3		4¾	35	6·3		9¼	72	12·1
	5¼	32	9·6		9	59	9·0		5	37	4·1		9½	74	11·6
	5½	34	2·4		9¼	61	4·3		5¼	39	1·9		9¾	76	11·1
	5¾	35	11·3		9½	62	14·8		5½	40	15·7		10	78	10·5
	6	37	4·6		9¾	64	9·3		5¾	42	13·5		10¼	80	10·0
	6¼	38	13·0		10	66	3·8		6	44	11·4		10½	82	9·4
	6½	40	5·8		10¼	67	14·3		6¼	46	9·2		10¾	84	8·9
	6¾	41	14·6		10½	69	8·8		6½	48	7·0		11	86	8·4
	7	43	7·5		10¾	71	3·3		6¾	50	4·8		11¼	88	7·8
	7¼	45	0·3		11	72	13·8		7	52	2·6		11½	90	7·3
	7½	46	9·2		11¼	74	8·3		7¼	54	0·4		11¾	92	6·8
	7¾	48	2·0		11½	76	2·8		7½	55	14·2		12	94	6·2
	8	49	10·8		11¾	77	13·3		7¾	57	12·0				
	8¼	51	3·7		12	79	7·8		8	59	9·8	2½	5	41	6·4
	8½	52	12·5						8¼	61	7·6		5¼	43	7.5
	8¾	54	5·4	2⅛	4½	29	14·5		8½	63	5·4		5½	45	8.6

T designates the thickness, B. the breadth.

T.	B.	Weight.		T.	B.	Weight.		T.	B.	Weight.		T.	B.	Weight.	
in.	in.	lbs	ozs.	in.	in.	lbs.	ozs.	in.	in.	lbs.	ozs.	in.	in.	lbs.	ozs.
2½	5¾	47	9·7	2⅝	7	60	13·7	2¾	8½	77	6·6	2⅞	10¼	97	9·6
	6	49	10·8		7¼	63	0·5		8¾	79	11·1		10½	99	15·7
	6¼	51	12·0		7½	65	3·2		9	81	15·5		10¾	102	5·7
	6½	53	13·1		7¾	67	6·0		9¼	84	3·9		11	104	11·8
	6¾	55	14·2		8	69	8·8		9½	86	8·4		11¼	107	1·9
	7	57	15·3		8¼	71	11·6		9¾	88	12·8		11½	109	8·0
	7¼	60	0·4		8½	73	14·3		10	91	1·2		11¾	111	14·1
	7½	62	1·6		8¾	76	1·1		10¼	93	5·7		12	114	4·2
	7¾	64	2·7		9	78	3·9		10½	95	10·1				
	8	66	3·8		9¼	80	6·7		10¾	97	14·5	3	6	59	9 8
	8¼	68	4·9		9½	82	9·4		11	100	3·0		6¼	62	1·6
	8½	70	6·0		9¾	84	12·2		11¼	102	7.4		6½	64	9·3
	8¾	72	7·2		10	86	15·0		11½	104	11·8		6¾	67	1·0
	9	74	8·3		10¼	89	1·8		11¾	107	0·3		7	69	8·8
	9¼	76	9·4		10½	91	4·6		12	109	4·7		7¼	72	0·5
	9½	78	10·5		10¾	93	7·3						7½	74	8·3
	9¾	80	11·6		11	95	10·1	2⅞	5¾	54	12·0		7¾	77	0·0
	10	82	12·8		11¼	97	12·9		6	57	·2·1		8	79	7·8
	10¼	84	13·9		11½	99	15·7		6¼	59	8·2		8¼	81	15·5
	10½	86	15·0		11¾	102	2·4		6½	61	14·2		8½	84	7·3
	10¾	89	0·1		12	104	5·2		6¾	64	4·3		8¾	86	15·0
	11	91	1·2						7	66	10·4		9	89	6·7
	11¼	93	2·4	2¾	5½	50	1 5		7¼	69	0·5		9¼	91	14·5
	11½	95	3·5		5¾	52	5·9		7½	71	6·6		9½	94	6·2
	11¾	97	4·6		6	54	10·3		7¾	73	12·7		9¾	96	14·0
	12	99	5·7		6¼	56	14·8		8	76	2·8		10	99	5·7
					6½	59	3·2		8¼	78	8·9		10¼	101	13.5
2⅝	5¼	45	10·3		6¾	61	7·6		8½	80	15.0		10½	104	5.2
	5½	47	13·0		7	63	12·1		8¾	83	5·0		10¾	106	13.0
	5¾	49	15·8		7¼	66	0·5		9	85	11·1		11	109	4.7
	6	52	2·6		7½	68	4·9		9¼	88	1·2		11¼	111	12.4
	6¼	54	5·4		7¾	70	9·4		9½	90	7·3		11½	114	4.2
	6½	56	8·1		8	72	13·8		9¾	92	13·4		11¾	116	11.9
	6¾	58	10·9		8¼	75	2·2		10	95	3·5		12	119	3.7

OBSERVATIONS ON TABLE OF FLAT IRON.

The weights here given are in pounds, ounces, and decimal parts, avoirdupois; and it will be seen, on inspecting the Table, that the first numbers in each page are those which apply to nut iron, and that the breadth increases by ¼ of an inch. The last numbers in each page show the weight of a square foot, according to the respective thickness of each bar. Hence the weight of any length of a bar of rectangular iron may be ascertained simply, as follows:

Rule.—Multiply the tabular weight, according to the thickness and breadth, by the number of feet in the bar, the product will be the weight required.

Example.—In a bar of iron whose thickness is 2¼ inches, the breadth 6½ inches, and the length 18 feet, what is the weight thereof?.

In the Table for 2¼ inches thick, and opposite 6½ inches, stand 48 lbs. 7 ozs.; being the weight of one lineal foot. Multiply this number by 18 feet, and we have as follows;

48 lbs. 7 ozs. × 18 = 871 lbs. 14 ozs.

17

· ELASTIC FORCE OF STEAM.

*Table of the Elastic Force of Steam, and corresponding Tempera-
ture of the Water with which it is in Contact.*

Pressure in pounds per sq. in.	Elastic force in Inches of Mercury.	Temperature Fahren't.	Volume of Steam compared with Vol. of Water.	Pressure in pounds per sq. in.	Elastic force in Inches of Mercury.	Temperature Fahren't.	Volume of Steam compared with Vol of Water'
14.7	30.00	212.0	1700	63	128.52	299.2	419
15	30.60	212.8	1669	64	130.50	300.3	443
16	32.64	216.3	1573	65	132.60	301.3	437
17	34.68	219.6	1488	66	134.64	302.4	431
18	36.72	222.7	1411	67	136.68	303.4	425
19	38.76	225.6	1343	68	138.72	304.4	419
20	40.80	229.5	1281	69	140.76	305.4	411
21	42.84	231.2	1225	70	142.80	306.4	408
22	44.88	233.8	1174	71	144.81	307.4	403
23	46.92	236.3	1127	72	146.88	308.4	398
24	48.96	238.7	1084	73	148.92	309.3	393
25	51.00	241.0	1044	74	150.96	310.3	388
26	53.04	243.3	1007	75	153.02	311.2	383
27	55.08	245.5	973	76	155.06	312.2	379
28	57.12	247.6	941	77	157.10	313.1	374
29	59.16	249.6	911	78	159.14	314.0	370
30	61.21	251.6	883	79	161.18	314.9	366
31	63.24	253.6	857	80	163.22	315.8	362
32	65.28	255.5	833	81	165.26	316.7	359
33	67.32	257.3	810	82	167.30	317.6	351
34	69.36	259.1	788	83	169.34	318.4	350
35	71.40	260.9	767	84	171.38	319.3	346
36	73.44	262.6	748	85	173.42	320.1	342
37	75.48	264.3	729	86	175.46	321.0	339
38	77.52	265.9	712	87	177.50	321.8	335
39	79.56	267.5	695	88	179.54	322.6	332
40	81.60	269.1	679	89	181.58	323.5	328
41	83.64	270.6	664	90	183.62	324.3	325
42	85.68	272.1	649	91	185.66	325.1	322
43	87.72	273.6	635	92	187.70	325.9	319
44	89.76	275.0	622	93	189.74	326.7	316
45	91.80	276.4	610	94	191.78	327.5	313
46	93.84	277.8	598	95	193.82	328.2	310
47	95.88	279.2	586	96	195.86	329.0	307
48	97.92	280.5	575	97	197.90	329.8	304
49	99.96	281.9	564	98	199.92	330.5	301
50	102.00	283.2	554	99	201.96	331.3	298
51	104.04	284.4	544	100	204.01	332.0	295
52	106.08	285.7	534	110	224.40	339.2	271
53	108.12	286.9	525	120	244.82	345.8	251
54	110.16	288.1	516	130	265.23	352.1	233
55	112.20	289.3	508	140	285.61	357.9	218
56	114.24	290.5	500	150	306.03	363.4	205
57	116.28	291.7	492	160	326.42	368.7	193
58	118.32	292.9	484	170	346.80	373.6	183
59	120.36	294.2	477	180	367.25	378.4	174
60	122.40	295.6	470	190	387.61	382.9	166
61	124.44	296.9	463	200	408.04	387.3	158
62	126.48	298.1	456				

Water holding impurities in solution tends to retard its attaining the aëriform
state, and so impairs the amount of its elastic force at an equal temperature.

Common water............. }	boiling point, 212° F. {	elastic force, 30 inches.
Sea water................. }	at 212 " {	" 23.05 "
Common water............. }	boiling point, 216° F. {	" 32.5 "
Sea water................. }	at 216 " {	" 24.6 "

PRODUCTION AND PROPERTIES OF STEAM.

When water in a vessel is subjected to the action of fire, it readily imbibes the heat or fluid principle of which the fire is the immediate cause, and sooner or later, according to the intensity of the heat, attains a temperature of 212° Fahrenheit. If at this point of temperature the water be not enclosed, but exposed to atmospheric pressure, ebullition will take place, and steam or vapor will ascend through the water, carrying with it the superabundant heat, or that which the water cannot under such circumstances of pressure absorb, to be retained and to indicate a higher temperature.

Water, in attaining the aeriform state, is thus uniformly confined to the same laws under every degree of pressure; but as the pressure is augmented, so is the indicated temperature proportionately elevated: hence the various densities of steam, and corresponding degrees of elastic force.

The preceding Table is peculiarly adapted for estimating the power of steam engines on the condensing principle, because in such the effective force of the steam is the difference between the total force and the resisting vapour retained in the condenser. The following Table is more adapted for estimating the effects of non-condensing engines, as, in such, the atmospheric pressure is not generally taken into account, engines of this principle being supposed to work in a medium; or, the atmospheric pressure on the boiler, to cause a greater density of steam, is equal to the resisting atmosphere which the effluent steam has to contend with on leaving the cylinder.

Table of the Elastic Force of Steam, the Pressure of the Atmosphere not being included.

Elastic Force in			Temperature in degrees of Fuhr.	Volume of Steam Water being 1.	Cubic in. of Water in a cubic foot of Steam.
Atmosphere.	lbs. square inch.	inch. of Mer.			
1.19	2.5	5.15	220	1496	1.11
1.22	3	6.18	222	1153	1.18
1.29	4	8.24	225	1366	1.25
1.36	5	10.3	228	1282	1.33
1.70	10	20.6	240	1041	1.64
2.04	15	30.9	251	883	1.93
2.38	20	41 2	260	767	2.23
2.72	25	51.5	263	678	2.52
3.06	30	61.8	275	609	2.81
3.40	35	72.1	282	553	3.09
3.74	40	82.4	288	506	3 38
4.08	45	92.7	294	465	3.66
4.42	50	103.0	299	435	3.93
4.76	55	113.3	304	407	4.20
5.10	60	123.6	309	382	4.48

Steam, independent of the heat indicated by an immersed thermometer, also contains heat that cannot be measured by any instrument at present known, and, in consequence of which, is termed latent or concealed heat; the only positive proof we have of its existence being that of incontestable results or effects produced on various bodies. Thus, if one part by weight of steam at 212° be mixed with nine parts of water at 62°, the result is water at 178.6°; therefore, each of the nine parts of water has received from the steam 116.6° of heat, and consequently the steam has diffused or given out $116.6 \times 9 = 1049.4 - 33.4 = 1016°$ of heat which it must have contained. Again, it is ascertained by experiment, that if one gallon of water be transformed into steam at 212°, and that steam allowed to mix with water at 52°, the whole will be raised to the boiling point, or 212°. From these and other experiments, it is ascertained that the latent heat in steam varies from 940°

to 1044°, the ratio of accumulation advancing from 212°, as the steam becomes more dense and of greater elastic force; hence the severity of a scald by steam to that by boiling water.

The rules formed by experimenters as corresponding with the results of their experiments on the elastic force of steam at given temperatures vary, but approximate so closely that the following rule, because of being simple, may in practice be taken in preference to any other.

Rule.—To the temperature of the steam in degrees of Fahrenheit, add 100, divide the sum by 177, and the 6th power of the quotient equals the force in inches of mercury.

Ex. Required the force of steam corresponding to a temperature of 312°.

$$312 + 100 \div 177 = 2.327^6 = 159 \text{ inches of mercury.}$$

But the Table is much better adapted to practical purposes, as the various results or effects are obtained simply by inspection.

CONSUMPTION OF COAL.

TABLE for finding the CONSUMPTION of COAL per Hour in Steamers either Paddle or Screw (the same Screw being used throughout.) at any Rate of Speed, the Consumption for a particular Rate being known. (At a given Amount of Coal, the Engineer may determine the most prudent Rate of Engine for reaching next coaling Port.)—*Engineer's and Contractor's Pocket Book, London.*

Speed.	Consumption of Coal	Speed.	Consumption of Coal.	Explanation.
3	.216	9	5.83	
3 1-2	.343	9 1-2	6.86	The speed for the consumption of a unit of coal is sup-
4	.512	10	8.00	tion of a unit of coal is sup-
4 1-2	.729	10 1-2	9.26	posed here to be 5, which may
5	1.000	11	10.65	be 5 miles or knots, or 5 times
5 1-2	1.331	11 1-2	12.15	any number of miles or knots;
6	1.728	12	13.82	then if 5 of such number of
6 1-2	2.197	12 1-2	15.61	miles require 1 unit of coal
7	2.744	13	17 58	per hour, 9 of such units will,
7 1-2	3.375	13 1-2	19.68	by the table, require 5.83 units
8	4.096	14	21.95	of coal, and 3 of them .216
8 1-2	4.910			units of coal.

It will be evident that this Table is calculated on the principle that the horse power varies very nearly as the cube of the speed; the enormous increase of consumption at increased velocities is in fact a little greater than that shown by the Table.

The advantages indicated above to be obtained at low velocities are evidently independent of those obtained at those velocities by using the steam expansively.

EVAPORATIVE POWER OF COAL AND RESULTS OF COKING.

Under the authority of an Act of the American Congress, approved Sept. 11, 1841, an extensive series of experiments was conducted by Prof. Johnson upon the evaporative power of several kinds of coal. The number of samples tried was 41, including 9 anthracites from Pennsylvania; 12 free-burning or semi-bituminous coals; 11 bituminous from Virginia; 6 foreign bituminous coals, viz. 1 from Sydney, Nova Scotia, sent by the Cunard Coal Mining Company; 1 of Pictou Coal, sent by the same; 1 of Scotch; 1 of Newcastle; 1 of Liverpool; and 1 of Pictou. From one to six trials were made on each sample, the average quantity used per trial being 978 lbs. The experiments occupied 114 days, during each of which continuous observations were made during 12 or 14 hours.

The coals were burnt under a steam boiler, fitted with apparatus for complete regulation, the supply of water and coals being determined both by weight and measure.

The standard adopted to measure the heating power of each kind of coal was the weight of water which a given weight of each evaporated from the temperature of 212° Fahr.

The following Table gives the results of five comparisons in each of which that coal which ranks the highest is stated as 1000, and the others in decimal parts of the integer.

Kinds of Coal.	Comparison 1.		Comparison 2.		Comparison 3.		Comparison 4.		Comparison 5.		
	Pounds of steam raised from water at 212° Fahr. by 1 lb. of fuel.	Relative evaporative power for equal weights of coal.	Pounds of steam from 212° produced by 1 cubic foot of each.	Relative evaporative power for equal bulks of coal.	Percentage of total waste in clinker and ashes.	Relative freedom from waste.	Time required to bring the boiler to steady action; in hours.	Relative rapidity of ignition.	Pounds of unburnt coke on the grate after each trial.	Relative completeness of combustion.	Weight in pounds to a cubic foot by experiment.
Anthracites :											
Atkinson and Templeman's	10.70	1.000	566.2	1.000	7.96	.633	0.99	.505	5.1	.725	52.92
Beaver Meadow (No. 5).	9.88	.923	556.1	.982	6.74	.748	2.42	2.07	6.12	.060	56.19
Bituminous and free burning :											
Newcastle .	8.66	.809	439.6	.776	5.68	.887	0.84	.595	10.7	.316	50.82
Pictou . . .	8.48	.792	417.9	.739	12.06	.418	0.95	.588	3.7	1.000	49.25
Liverpool .	7.84	.733	375.4	.663	5.04	1.000	0.86	.581	11 1	.333	47.88
Cannelton, (In)	7.34	.686	348.9	.616	5.12	.984	0.50	1.000	6.4	.578	47.65
Scotch . .	6.95	.619	353.8	.625	10.10	.499	0.96	.521	5.7	.649	51.05
Dry pine wood.	4.69	.436	98 6	.175	0.307	16.417					

The same report states some results of coke-burning, from which it appears that by burning in uncovered heaps, and only covering up the ignited mass when flame ceases to be emitted (as in many of the iron works of Great Britain, France, &c.), the loss in weight at Plymouth has been found to be 17 per cent.; at Penn-y-darran, 20 per cent.; and at Dowlais (where it may be presumed the abundance of coal admits of an uneconomical management), 34 per cent. By coking in stacks, or well covered heaps of coal from 10 to 15 ft. diameter, as followed in Staffordshire, highly bituminous coals lose from 50 to 55 pr. ct. weight, and those of a drier nature from 35 to 40.

By coking in close ovens, a coal which, in an uncovered heap, yields only 45 to 50 per cent., yields 69 per cent. In the close oven the *gain* in *bulk* is from 22 to 23 per cent.; and while highly bituminous coals yield only 40 to 45 per cent. in open heaps, and actually *lose* in *bulk*, they yield in close ovens from 65 to 66 per cent., and *gain* in *bulk*. By coking in gas retorts, the Deane Coal of Cumberland gains nearly 30 per cent. in bulk, and loses in weight 25 per cent. Carlisle coal nearly the same. Cannel and Cardiff coals gain 30 per cent. in bulk, and lose 36.5 in weight. Bewick's Wallsend loses 30, and Russell's Wallsend, 30.7 per cent, by the same process.

17*

POWER OF STEAM.

Mr. Tredgold gives the following Table, which will show how the power of the steam as it issues from the boiler, is distributed.

IN A NON-CONDENSING ENGINE.

Let the pressure on the boiler be	10.000
Force required to produce motion of the steam in the cylinder will be 0.069	
Loss by cooling in the cylinder and pipes 0.160	
Loss by friction of the piston and waste 2.000	
Force required to expel the steam into the atmosphere 0.069	
Force expended in opening the valves, and friction of the various parts 0.622	
Loss by the steam being cut off before the end of the stroke 1.000	
Amount of deductions ——— 3.920	
Effective pressure............ 6.080	

IN A CONDENSING ENGINE.

Let the pressure on the boiler be	10.000
Force required to produce motion of the steam in the cylinder...... 0.070	
Loss by cooling in the cylinder and pipes......................... 0.160	
Loss by friction of the piston and waste........................... 1.250	
Force required to expel the steam through the passages............ 0.070	
Force required to open and close the valves, raise the injection	
water, and overcome the friction of the axes.................... 0.630	
Loss by the steam being cut off before the end of the stroke........ 1.000	
Power required to work the air-pump............................ 0.500	
Amount of deductions ——— 3.680	
Effective pressure............ 6.320	

If we now suppose a cylinder whose diameter is 24 inches, the area of this cylinder and consequently the area of the piston in square inches, will be,

$$24^2 \times .7854 = 452.39$$

Let us also make the supposition that steam is admitted into the cylinder of such power as exerts an effective pressure on the piston of 12 lbs. to the square inch; therefore, $452.39 \times 12 = 5428.68$ lbs., the whole force with which the piston is pressed. If we now suppose that the length of the stroke is five feet, and the engine makes 44 single or 22 double strokes in a minute, then the piston will move through a space of $22 \times 5 \times 2 = 220$ feet in a minute; the power of the engine being equivalent to a weight of 5428 lbs. raised through 220 feet in a minute.

This is the most certain measure of the power of a steam engine. It is usual, however, to estimate the effect as equivalent to the power of so many horses. This method, however simple and natural it may appear, is yet, from differences of opinion as to the power of a horse, not very accurate; and its employment in calculation can only be accounted for on the ground, that when steam engines were first employed to drive machinery, they were substituted instead of horses; and it became thus necessary to estimate what size of a steam engine would give a power equal to so many horses.

There are various opinions as to the power of a horse. According to Smeaton, a horse will raise 22,916 lbs. one foot high in a minute. Desaguliers makes the number 27,500; and Watt makes it larger still, that is 33,000. There is reason to believe that even this number is too small, and that we may add at least 11,000 to it, which gives 44,000 lbs. raised one foot high per minute.—*Grier.*

RULES AND TABLES

FOR

GAUGING, ULLAGING, &c.

GAUGING OF CASKS.

In taking the dimensions of a Cask it must be carefully observed : 1st, That the bung-hole be in the middle of the cask ; 2d, That the bung-stave, and the stave opposite to the bung-hole, are both regular and even within ; 3d, That the heads of the Cask are equal, and truly circular ; if so, the distance between the inside of the chime to the outside of the opposite stave will be the head diameter within the Cask, very near.

RULE. — Take, in inches, the *inside* diameters of a Cask at the Head and the Bung, and also the Length ; subtract the head-diameter from the bung-diameter, and note the difference.

If the measure of the Cask is taken outside, with callipers, from head to head, then a deduction must be made of from 1 to 2 inches for the thickness of the heads, according to the size of the Cask.

1 *If the staves of the Cask, between the bung and the head, are considerably curved,* (the shape of a Pipe), multiply the difference between the bung and head, by .7.

2 *If the staves be of a medium curve,* (the shape of a Molasses Hogshead), multiply the difference by .65.

3 *If the staves curve very little,* (less than a Molasses Hogshead), multiply the difference by .6.

4. *If the staves are nearly straight,* (almost a Cylinder), multiply the difference by .55.

5. Add the product, in each case, to the head-diameter ; the sum will be a mean diameter, and thus the Cask is reduced to a cylinder.

6. Multiply the *mean* diameter by itself, and then by the length, and multiply if for Wine gallons, by .0034. The difference of dividing by 294 (the usual method), and multiplying by .0034 (the most expeditious method), is less than 500ths of a gallon in 100 gallons.

EXAMPLE.

Supposing the Head-Diameter of a Cask to be 24 inches, the Bung-Diameter 32 inches, and the Length of Cask 40 inches; What is the content in Wine Gallons ?   1st *variety.*

Bung-Diameter,	32	brought up	876.16
Head-Diameter,	24	Length,	40
Difference,	8		35046.40
Multiplier,	.7		.0034
	5.6		14018560
Head-Diam.,	24		10513920
multiply	29.6		119.157760
by	29.6		
carry up Square,	876.16	*Ans.* 119 galls. 1 pint.	

To obtain the contents of a similar Cask in Ale Gallons, multiply 35046.40 by .002785, and we get 97.6042, (or 97 gallons 5 pints.)

GAUGING OF CASKS IN IMPERIAL (BRITISH) GALLONS. AND ALSO IN UNITED STATES GALLONS.

Having ascertained the *variety* of the Cask, and its *interior* dimensions, the following Table will facilitate the calculation of its capacity.

Table of the Capacities of Casks, whose Bung Diameters and Lengths are 1 or Unity.

H.	1st Var.	2d Var.	3d Var.	4th Var.	H.	1st Var.	2d Var.	3d Var.	4th Var.
.50	.0021244	.0020300	.0017704	.0016523	.76	.0024337	.0024120	.0022343	.0022071
.51	.0021310	.0020433	.0017847	.0016713	.77	.0024482	.0024282	.0022560	.0022310
.52	.0021437	.0020567	.0017993	.0016905	.78	.0024628	.0024445	.0022780	.0022551
.53	.0021536	.0020702	.0018141	.0017098	.79	.0024777	.0024610	.0023002	.0022794
.54	.0021637	.0020838	.0018293	.0017294	.80	.0024927	.0024776	.0023227	.0023038
.55	.0021740	.0020975	.0018447	.0017491	.81	.0025079	.0024942	.0023455	.0023285
.56	.0021845	.0021114	.0018604	.0017690	.82	.0025233	.0025110	.0023686	.0023533
.57	.0021951	.0021253	.0018761	.0017891	.83	.0025388	.0025279	.0023920	.0023783
.58	.0022060	.0021394	.0018927	.0018094	.84	.0025546	.0025449	.0024156	.0024035
.59	.0022170	.0021536	.0019093	.0018299	.85	.0025706	.0025621	.0024396	.0024289
.60	.0022283	.0021679	.0019261	.0018506	.86	.0025867	.0025793	.0024638	.0024545
.61	.0022397	.0021823	.0019433	.0018715	.87	.0026030	.0025967	.0024883	.0024803
.62	.0022513	.0021968	.0019607	.0018925	.88	.0026196	.0026141	.0025131	.0025063
.63	.0022631	.0022114	.0019784	.0019138	.89	.0026363	.0026317	.0025381	.0025324
.64	.0022751	.0022262	.0019964	.0019352	.90	.0026532	.0026494	.0025635	.0025588
.65	.0022873	.0022410	.0020117	.0019568	.91	.0026703	.0026672	.0025891	.0025853
.66	.0022997	.0022560	.0020332	.0019786	.92	.0026875	.0026851	.0026150	.0026120
.67	.0023122	.0022711	.0020521	.0020006	.93	.0027050	.0027032	.0026412	.0026389
.68	.0023250	.0022863	.0020712	.0020228	.94	.0027227	.0027213	.0026677	.0026660
.69	.0023379	.0023016	.0020906	.0020452	.95	.0027405	.0027396	.0026915	.0026933
.70	.0023510	.0023170	.0021103	.0020679	.96	.0027585	.0027579	.0027215	.0027208
.71	.0023643	.0023326	.0021302	.0020905	.97	.0027768	.0027761	.0027489	.0027484
.72	.0023778	.0023482	.0021505	.0021135	.98	.0027952	.0027950	.0027765	.0027763
.73	.0023915	.0023640	.0021710	.0021366	.99	.0028138	.0028137	.0028044	.0028013
.74	.0024054	.0023799	.0021918	.0021599	1.00	.0028326	.0028326	.0028326	.0028326
.75	.0024195	.0023959	.0022129	.0021834					

Divide the head by the bung diameter, and opposite the quotient in the column H, and under its proper variety, is the tabular number for unity. Multiply the tabular number by the square of the bung diameter of the given cask, and by its length, the product equals its capacity in Imperial gallons.

Required the number of Gallons in a Cask, (*1st variety*,) 24 inches head diameter, 32 bung diameter, and 40 inches in length?

32) 24.0 (.75 see Table for tabular No.
.0024195 tabular No. for unity.
32 × 32 is 1024 square of bung diam.

```
        96780
        48390
        24195
      2.4775680
         40 Inches long.
     99.1027200 Imperial Gallons.
            1.2
     1982054400
      991027200
    118.92326400 United States Gallons.
```

NOTE. — Multiplying Imperial gallons by one & two-tenths (1.2) will convert them into U. S. gallons; and U. S. gallons multiplied by ·833 equal Imperial gallons.

TO ULLAGE, OR FIND THE CONTENTS IN GALLONS OF A CASK PARTLY FILLED.

To find the contents of the occupied part of a lying cask in gallons.

RULE.—Divide the depth of the liquid, or wet inches, by the bung diameter, and if the quotient is under .5 deduct from the quotient *one-fourth* of what it is less than .5, and multiply the remainder, by the whole capacity of the cask, this product will be the number of gallons in the cask. But if the quotient exceeds .5, add *one-fourth* of that excess to the quotient, and multiply the sum, by the whole capacity of the cask, this product will be the number of gallons.

EXAMPLE I.—Suppose the bung-diameter of a cask, on its bilge, is 32 inches, and the whole contents of the cask 118.80 U. S. standard gallons; required the ullage of 15 wet inches.

$$32) \; 15.00 \; (.46875 \quad .5 - .46875 = .03125 \div 4 = .0078125 \quad .46875 -$$
$$.0078125 = .4609375 \times 118.80 = 54.759375 \; \text{U. S. Gallons.}$$

EXAMPLE II.—Required the ullage of 17 wet inches in a cask of the above capacity ?

$$32) \; 17.00 \; (.53125 - .5 = .03125 \div 4 = .0078125 + .53125 = .5390625$$
$$\times 118.80 = 64.040625 \; \text{U. S. Gallons.}$$

PROOF — $64 \cdot 040625 + 54 \cdot 759375 = 118 \cdot 80$ *gallons.*

To find the ullage of a filled part of a standing Cask, in gallons.

RULE.—Divide the depth of the liquid, or wet inches, by the length of the cask; then, if the quotient is less than .5, deduct from the quotient *one-tenth* of what it is less than .5 and multiply the remainder, by the whole capacity of the cask, this product will be the number of gallons. But if the quotient exceeds .5, add *one-tenth* of that excess to the quotient, and multiply the sum, by the whole capacity of the cask, this product will be the ullage, or contents in U. S. standard gallons.

EXAMPLE.—Suppose a cask, 40 inches in length, and the capacity 118.80 gallons, as above: required the ullage of 21 wet inches ?

$$40) \; 21.000 \; (.525 - .5 = .025 \div 10 = .0025 + .525 = .5275 \times 118.80$$
$$= 62.667 \; \text{U. S. Gallons.}$$

NOTE.—Formerly the British Wine and Ale Gallon measures were similar to those now used in the United States and British Colonies.

The following Tables exhibit the comparative value between the United States and the present British measures.

U. S. measure for wine, spirits, &c.	British (Im.) measure. galls. qts. pts. gills.				U. S. measure for ale and beer.	British (Im.) measure. galls. qts. pts. gills.			
42 galls. = 1 tierce,	= 34	3	1	3	9 galls. = 1 firkin,	= 9	0	1	1
63 = 1 hogsh.	= 52	1	1	3	36 = 1 barrel,	= 36	2	0	3
126 = 1 pipe,	= 104	3	1	3	54 = 1 hogsh.	= 54	3	1	1
252 = 1 tun,	= 209	3	1	2	108 = 1 butt,	= 109	3	0	3

To convert Imperial Gallons into United States Wine Gallons multiply the imperial by 1·2. To convert U. S. Gallons into Imperial multiply the U. States Wine gallons by ·833.

51 U. S. Ale Gallons equal 60 Imperial Gallons, therefore to convert one into other add or deduct 1-60th.

PLOUGHING.

Table showing the distance Travelled by a horse in Ploughing an Acre of Land; also, the quantity of Land worked in a Day, at the rate of 16 and 18 miles per day of 9 hours.

B'dth of Furrow slice.	Space travel-led in Plough-ing an Acre.	Extent Ploughed per Day.		B'dth of Furrow slice	Space travel-led in Plough-ing an Acre.	Extent Ploughed per Day.	
Inches.	Miles.	18 Miles.	16 Miles.	Inches.	Miles.	18 Miles.	16 Miles.
7	14 1-2	1 1-4	1 1-8	14	7	2 1-2	2 1-4
8	12 1-2	1 1-2	1 1-4	15	6 1-2	2 3-4	2 2-5
9	11	1 3-5	1 1-2	16	6 1-6	2 9-10	2 3-5
10	9 9-10	1 4-5	1 3-5	17	5 3-4	3 1-10	2 3-4
11	9	2	1 3-4	18	5 1-2	3 1-4	2 9-10
12	8 1-4	2 1-5	1 9-10	19	5 1-4	3 1-2	3 1-10
13	7 1-2	2 1-3	2 1-10	20	4 9-10	3 1-5	3 1-4

PLANTING.

Table showing the number of Plants required for one Acre of Land, from one Foot to Twenty-one Feet *distance* from Plant to Plant.

Feet Distance.	No. of Hills.	Feet Distance.	No. of Hills.	Feet Distance.	No. of Hills.	Feet Distance.	No. of Hills.	Feet Distance.	No. of Hills
1	43,560	4	2,722	7	889	10	436	17	151
1½	19,360	4½	2,151	7½	775	10½	361	18	135
2	10,890	5	1,742	8	680	12	302	20	108
2½	6,969	5½	1,440	8½	602	14	223	21	99
3	4,840	6	1,210	9	538	15	193	25	69
3½	3,556	6½	1,031	9½	482	16	171	30	48

WEIGHT OF A CORD OF WOOD.

Table of the Weight of a Cord of different kinds of Dry Wood, and the comparative value per Cord.

A Cord of Hickory,	- - 4469 pounds,	- - Carbon	- - 100
" Maple,	- - - 2863 "	- - "	- - 54
" White Birch,	- 2369 "	- - "	- - 48
" " Beech,	- 3236 "	- - "	- - 65
" " Ash,	- - 3450 "	- - "	- - 77
" Pitch Pine,	- - 1904 "	- - "	- - 43
" White Pine,	- 1868 "	- - "	- - 42
" Lombardy Poplar 1774	"	- - "	- - 40
" White Oak	- - 3821 "	- - "	- - 81
" Yellow Oak,	- 2919 "	- - "	- - 60
" Red Oak,	- - 3254 "	- - "	- - 69

NOTE.— Nearly one half of the weight of a growing Oak tree consists of sap. Ordinary Dry Wood contains about one-fourth of its weight in water.

CHARCOAL.

Oak, Maple, Beech, and Chestnut make the best quality. Between 15 and 17 per cent. of coal can be obtained when the wood is properly burned. A bushel of coal from hard wood weighs between 29 and 31 lbs., and from from pine between 28 and 30 lbs.

TINMAN'S TWELVE POUND BILL,

OR BILL OF DAY'S WORK

OF PLAIN TIN WARE.

No. of Articles for Day's Work.	12 lb.	No. of Articles for Day's Work.	12 lb.
16 Sixteen quart Large Dish Kettles,	84	36 Round Toast Pans,	18
10 Water Pots,	75	40 Quart Covered Pails,	18
18 Twelve quart Pails,	67	36 Round Flat Bottom Tea Pots,	18
18 Large Dish Kettles,	67	72 Second Size Horn,	18
20 Foot Stoves,	67	48 Sailor Pots,	18
24 Ten quart Pails,	58	36 Quart Lamp Fillers,	18
24 Ten quart Pans,	68	36 Water Ladles,	18
18 Gallon Coffee Pots,	58	36 Sugar Scoops,	18
18 Six quart Covered Pails,	58	36 Milk Strainers,	18
18 Large Sauce Pans,	58	72 Quart Measures,	14
24 Gallon Measures,	39	48 Large Skimmers,	14
30 Six quart Pails,	39	72 Quart Funnels,	14
36 Common Size Milk Pans,	39	72 Small Horns,	14
20 Large Wash Bowls,	39	72 Basins,	12
20 Lanterns,	39	144 Quart Scollops,	12
24 Small Dish Kettles, six qt.	39	144 Quart Grease Pans,	12
20 Cullenders,	39	60 Round Handled Dippers,	12
24 Three quart Coffee Pots,	39	120 Half Square Pans,	10
24 Large Pudding Bags,	39	84 Half Sheet Funnels,	10
24 Roasters,	39	72 Half Sheet Dippers,	10
40 Lantern Pans,	36	120 Half Sheet Scollops,	10
24 Two quart Coffee Pots,	34	96 Pint Funnels,	8
20 Three qt. Covered Pails,	34	84 Pint Measures,	8
24 Small Wash Bowls,	34	96 Pint Cups,	8
24 Small Sauce Pans,	34	168 Pint Scollops,	8
30 Half gallon Measures,	25	48 Flour Boxes,	8
48 Half gallon Pans,	25	96 Half Pint Measures,	5
24 Half gallon Dippers,	25	108 Half Pint Cups,	5
36 Half gallon Funnels,	25	96 Half Pint Dippers,	5
30 Three pint Coffee Pots,	25	120 Half Pint Funnels,	5
24 Two quart Covered Pails,	25	96 Gill Measures,	5
86 Large Blow Horns,	25	48 Basters,	5
36 Three quart Pails,	25	96 Small Skimmers,	5
48 Round Pans,	18	124 Flat Candlesticks,	5
100 Square Pans,	18	120 Needle Cases,	5
108 Scollop Pie Pans,	18	84 Pepper Boxes,	5
48 Sausage Horns,	18	120 Hearts,	3
86 Quart Coffee Pots,	18	144 Rounds,	3
48 Square Toast Pans,	18	98 Rattle Boxes,	3

[The 6 Pound Bill is one-half of the 12 Pound Bill.]

No. of Articles for Day's Work.	12 lb.	No. of Articles for Day's Work.	12 lb.
12 Six quart Coffee Boilers,	1.00	24 Nine quart Pans,.......	
12 Five quart Coffee Boilers,	83	16 Twelve qt. Pans, handles,	80
12 Four quart Coffee Boilers,	67	20 Seven qt. Pans, handles,	50
12 Three qt. Coffee Boilers,	50	36 Five quart Straight Pans,	
12 Two quart Coffee Boilers,	42	40 Two quart Straight Pans,	
12 Six quart Coffee Pots,...	88	48 Three pint Straight Pans,	
12 Five quart Pots,........	75	20 Handled Wash Boards,..	39
12 Large Dutch Buckets,...		18 Twelve qt. Dish Kettles,.	67
12 Small Dutch Buckets,...		18 Ten qt. Dish Kettles,....	58
12 Small Water Pots,......		24 Four qt. Dish Kettles,...	39
12 Ten quart Covered Pails,	84	40 Three pint Dish Kettles,.	18
18 Five quart Covered Pails,	50	Twelve qt. Cov. Buckets,	1.00
26 Three pint Covered Pails,	20	Oak Leaf Cake Cutters,..	10
30 One pint Covered Pails,..	14	One quart Tea Pots,.....	84
30 Five quart Open Pails,..	50	One gallon Fluid Cans,..	
32 Gall. Open Pails,.......		Half gallon Fluid Cans,..	
40 Three Pint Open Pails,..			

1.—WEIGHTS OF IRON WIRE PER 20 FEET.

No. 0...5 lbs.		No. 6...1 lb. 14 ozs.		No. 12...9 ozs.	
No. 1...4 lbs. 2 ozs.		No. 7...1 lb. 10 ozs.		No. 13...6 ozs.	
No. 2...3 lbs. 8 ozs.		No. 8...1 lb. 7 ozs.		No. 14...5 ozs.	
No. 3...2 lbs. 15 ozs.		No. 9...1 lb. 2 ozs.		No. 15...$4\frac{1}{2}$ ozs.	
No. 4...2 lbs. 8 ozs.		No. 10...—— 14 ozs.		No 16...$3\frac{1}{2}$ ozs.	
No. 5...2 lbs. 5 ozs.		No. 11...—— 10 ozs.		No. 17...3 ozs.	

2.—WEIGHT OF IRON WIRE PER LINEAL ROD.

Nos.	Diameter in 1-100 of an Inch.	Weight per Lineal Rod.	Nos.	Diameter in 1-100 of an Inch.	Weight per Lineal Rod.
1	.32	4 lbs. 2 ozs.	8	.18	1 lb. 4 ozs.
2	.30	3 " 10 "	9	.16	1 " 0 "
3	.27	2 " 15 "	10	.15	0 " 14 "
4	.25	2 " 8 "	11	.13	0 " 10 "
5	.24	2 " 5 "	12	.12	0 " 9 "
6	.22	1 " 15 "	13	.10	0 " 6 "
7	.20	1 " 9 "			

ERRATA.

Page 35.—To find the Solidity of a Pyramid or Cone.

Rule.—Multiply the area of the base by the height, and one-third of the product will be the solid content.

Example.—Required the solid content in inches of a Cone or Pyramid, the diameter of the base being 8 inches, and perpendicular height 18 inches?

$8 \times 8 = 64 \times .7851 \times 18 = 904.7808 \div 3 = 301.5936$ inches $\div 231 = 1$ gall. $1\frac{1}{4}$ qts.

Page 92 *No.* 35.—For Tin 64 lbs Copper 1 lb. *read* Copper 64 lbs. Tin 1 lb.